Human Diet and Nutrition in Biocultural Perspective

Studies of the Biosocial Society

General Editor: **Catherine Panter-Brick,** Professor of Anthropology, Harvard University

The Biosocial Society is an international academic society engaged in fostering understanding of human biological and social diversity. It draws its membership from a wide range of academic disciplines, particularly those engaged in "boundary disciplines" at the intersection between the natural and social sciences, such as biocultural anthropology, medical sociology, demography, social medicine, the history of science and bioethics. The aim of this series is to promote interdisciplinary research on how biology and society interact to shape human experience and to serve as advanced texts for undergraduate and postgraduate students.

Volume 1
Race, Ethnicity, and Nation
Perspectives from Kinship and Genetics
Edited by Peter Wade

Volume 2
Health, Risk, and Adversity
Edited by Catherine Panter-Brick and Agustín Fuentes

Volume 3
Substitute Parents
Biological and Social Perspective on Alloparenting in Human Societies
Edited by Gillian Bentley and Ruth Mace

Volume 4
Centralizing Fieldwork
Critical Perspectives from Primatology, Biological and Social Anthropology
Edited by Jeremy MacClancy and Agustín Fuentes

Volume 5
Human Diet and Nutrition in Biocultural Perspective
Past Meets Present
Edited by Tina Moffat and Tracy Prowse

Human Diet and Nutrition in Biocultural Perspective

Past Meets Present

● ● ●

Edited by Tina Moffat and Tracy Prowse

Berghahn Books
New York • Oxford

*This book is in memory of Shelley Saunders,
our mentor and dear friend.*

First published in 2010 by
Berghahn Books
www.berghahnbooks.com

©2010 Tina Moffat and Tracy Prowse

All rights reserved. Except for the quotation of short passages
for the purposes of criticism and review, no part of this book
may be reproduced in any form or by any means, electronic or
mechanical, including photocopying, recording, or any information
storage and retrieval system now known or to be invented,
without written permission of the publisher.

Library of Congress Cataloging-in-Publication Data

Human diet and nutrition in biocultural perspective : past meets present / edited by Tina Moffat and Tracy Prowse.
  p. cm. — (Studies of the biosocial society)
Includes bibliographical references and index.
ISBN 978-1-84545-765-5
 1. Food habits. 2. Diet. 3. Human nutrition. I. Moffat, Tina, 1967– II. Prowse, Tracy Lynn.
GT2850.H86 2010
394.1'2—dc22
                                                                2010023854

British Library Cataloguing in Publication Data

A catalogue record for this book is available from the British Library

Printed in the United States on acid-free paper.

ISBN: 978-1-84545-765-5 Hardback

Contents

● ● ●

List of Figures	vii
List of Tables	x
List of Boxes	xii
Introduction. A Biocultural Approach to Human Diet and Nutrition T. Moffat and T. Prowse	1

Evolutionary Perspectives on Nutrition

1. What Did Humans Evolve to Eat? Metabolic Implications of Major Trends in Hominid Evolution W. R. Leonard, M. L. Robertson, and J. J. Snodgrass	13
2. Child Growth among Southern African Foragers in the Past S. Pfeiffer and L. Harrington	35
3. Infant and Young Child Feeding in Human Evolution D. W. Sellen	57

Breastfeeding and Beyond: Nutrition throughout the Life Course

4. The Use of Stable Isotope Analysis to Determine Infant and Young Child Feeding Patterns T. L. Dupras	89
5. A Community in Transition: Deconstructing Breastfeeding Trends in Gibraltar, 1955–1996 L. A. Sawchuk, E. K. Bryce, and S. D. A. Burke	109

Food Insecurity and Malnutrition

6. Dietary Diversity, Dietary Transitions, and Childhood Nutrition in Nepal: Questions of Methodology and Practice   133
T. Moffat and E. Finnis

7. Responses to a Food Crisis and Child Malnutrition in the Nigerien Sahel   152
R. E. Casiday, K. R. Hampshire, C. Panter-Brick, and K. Kilpatrick

Nutritional Factors in Growth and Disease

8. Growth, Morbidity, and Mortality in Antiquity: A Case Study from Imperial Rome   173
T. Prowse, S. Saunders, C. Fitzgerald, L. Bondioli, and R. Macchiarelli

9. Examining Nutritional Aspects of Bone Loss and Fragility across the Life Course in Bioarchaeology   197
S. C. Agarwal and B. Glencross

10. Obesity: An Emerging Epidemic—Temporal Trends in North America   223
P. T. Katzmarzyk

Conclusion. Diet and Nutrition in Biocultural Perspective: Back to the Future   241
T. Prowse and T. Moffat

Contributors   252

Glossary   258

Index   264

List of Figures

● ● ●

Figure 1.1. Log-Log plots of resting metabolic rate (RMR; kcal/day) versus body weight (kg) for 51 species of terrestrial mammals and brain weight (BW; g) versus RMR (kcal/day) for humans, 35 other primate species, and 22 nonprimate mammalian species   15

Figure 1.2. Plots of diet quality (DQ) versus log-body mass for 33 primate species and relative brain size versus relative diet quality for 31 primate species   17

Figure 1.3. Patterns of physical growth in stature (cm) and body fatness (as sum of triceps and subscapular skinfolds, mm) in girls of Tsimane' of lowland Bolivia   23

Figure 1.4. Total daily energy expenditure (TDEE; kcal/day) versus body weight (kg) for adult men and women of industrial and subsistence-level populations   28

Figure 2.1. The areas from which Later Stone Age skeletons have been recovered, within fynbos (striped), forest (solid), and savanna (stippled) ecosystems along the southern coast of South Africa   43

Figure 2.2. Basicranial size for age in Later Stone Age juveniles and three comparative groups   45

Figure 2.3. Femur length in Later Stone Age (LSA) juveniles expressed as a percentage of the mean value for femur length in LSA adults, plotted to compare with the percentage of adult length achieved in Denver growth study participants   47

Figure 2.4. Residual difference in percentage of adult femur length achieved by Later Stone Age juveniles (interpolated cohort means) and five other groups, each expressed relative to predicted values taken from the Denver growth study   49

Figure 3.1. Juvenile feeding ecology in nonhuman primates — 62

Figure 3.2. Evolved template for human infant feeding — 67

Figure 3.3. Coevolutionary forces influencing infant feeding among human ancestors — 71

Figure 4.1. Map of Egypt showing location of the Dakhleh Oasis and the ancient site of Kellis — 93

Figure 4.2. Excavation map from the Kellis 2 cemetery in the Dakhleh Oasis, Egypt — 95

Figure 4.3. Carbon and nitrogen mean isotopic values for infants from the Kellis 2 cemetery and enamel oxygen and carbon mean isotopic values of teeth from individuals from the Kellis 2 cemetery — 97

Figure 4.4. Age of formation for the mandibular deciduous and permanent dentition and dentin nitrogen and carbon isotopic values of teeth from individuals from the Kellis 2 cemetery — 100

Figure 5.1. Health in Gibraltar, 1960–1993: Infant mortality rate and life expectancy by sex — 115

Figure 5.2. Interior of the Maternity and Child Welfare Centre: Gibraltar — 120

Figure 5.3. Breastfeeding rates in Gibraltar: 1955–1996, by socioeconomic status — 125

Figure 6.1. Food type consumed on recall day by rural children by age group — 140

Figure 6.2. Food type consumed on recall day for urban sample by age group — 141

Figure 6.3. Frequency of Dietary Diversity Score (DDS) for rural and urban samples — 142

Figure 7.1. Affala, Niger, October 2005. Families participating in the Concern Worldwide emergency nutrition program queue to receive their "family ration" of grain, pulses, and oil. — 154

Figure 7.2. Plumpy'nut, a peanut-based therapeutic food used for treating infants and children with severe acute malnutrition, is provided ready-for-use in small sachets. — 160

Figure 8.1. Map of Portus and Isola Sacra — 175

Figure 8.2. Maximum prevalence by month (MAP) and smoothed curve, suggesting the shape of the "true" prevalence distribution (SMAP) — 183

Figure 8.3. Isola Sacra vs. Belleville Diaphyseal Lengths — 184

Figure 8.4. Scatter plot of the Isola Sacra rib data (n=37), showing $\delta^{15}N$ and $\delta^{13}C$ versus estimated age-at-death   186

Figure 9.1. Key determinants of bone fragility   199

Figure 9.2. Developmental influences on bone fragility from a lifecycle perspective   201

Figure 9.3. Basic model of bone fragility in a multilevel framework   205

Figure 9.4. High resolution pQCT images of age-related changes in trabecular architecture in fourth lumbar vertebrae from Neolithic archaeological skeletons from Çatalhöyük, Turkey   211

Figure 10.1. Estimated average body mass index in males across the evolution of hominids from *Australopithecus afarensis* to contemporary *Homo sapiens*   227

Figure 10.2. Prevalence of self-reported obesity (BMI ≥ 30 kg/m^2) in the United States and Canada, 1994 and 2005   229

Figure 10.3. Temporal trends in the prevalence of measured obesity (BMI ≥ 30 kg/m^2) among adults in Mexico from the 1992–93 National Survey of Chronic Diseases and the 2000 Mexican National Health Survey   233

List of Tables

● ● ●

Table 1.1. Geological ages (millions of years ago), brain size (cm^3), estimated male and female body weights (kg), and estimated home range sizes (hectares) for selected fossil hominid species 18

Table 1.2. Body weight (kg), brain weight (g), percent body fat (%), resting metabolic rate (RMR; kcal/day), and percent of RMR allocated to brain metabolism (BrMet, %) for humans from birth to adulthood 22

Table 1.3. Percentage of US Adults (20 years and older) who are obese (BMI ≥ 30 kg/m^2), 1960 to 2004 24

Table 1.4. Dietary energy (kcal/day) and macronutrient (percent of dietary energy) intakes among US adults (20 years ad older), 1971 to 2000 25

Table 1.5. Comparison of body weight (kg), total daily energy expenditure (TDEE; kcal/day), resting metabolic rate (RMR; kcal/day), and Physical Activity Level (PAL) of subsistence-level and industrial societies 27

Table 2.1. Maximum width (in millimeters) of the basiocciput in Later Stone Age juveniles and three other groups 44

Table 2.2. Estimated age at death and femur lengths for the Later Stone Age sample 46

Table 2.3. Comparative groups included in the analysis of linear growth 50

Table 3.1. Terms used to describe aspects of lactation biology 58

Table 3.2. Relationships between life history and lactation biology 60

Table 3.3. Phylogenetic relationships of great ape species and average values for selected life history parameters 65

Table 3.4. Evolutionary classification of characteristics of human lactation biology 66

Table 4.1. Number of individuals in each age category from the Kellis 2 cemetery   96

Table 5.1. Proportion of infants breastfed in Gibraltar by cohort   117

Table 5.2. Proportion of infants breastfed in Gibraltar by socioeconomic status (SES) and cohort   119

Table 6.1. Ten standard food groups for calculating Dietary Diversity Score (DDS)   139

Table 6.2. Comparison of DDS for urban and rural samples   143

Table 7.1. International frameworks and indicators of a food crisis   156

Table 7.2. Child health and nutrition indicators for Nigerien children (6–59 months)   157

Table 10.1. Classification of body weight status in adults using the body mass index (BMI)   224

Table 10.2. International classification of body weight status among children and youth using the body mass index   226

Table 10.3. Temporal trends in the prevalence of measured obesity (BMI ≥ 30 kg/m^2) in Canada and the United States (1972–2005)   230

Table 10.4. Prevalence of abdominal obesity (waist circumference > 88 cm in women; > 102 cm in men) among participants in the 1981 Canada Fitness Survey, 1988 Campbell's Survey of Well-being, and the 1986–1992 Canadian Heart Health Surveys   231

Table 10.5. Summary relative risk estimates and population attributable risks (PAR%) for obesity in Canada   234

Table 10.6. Direct and indirect costs (CAD millions) of major chronic diseases associated with obesity in Canada, 2001   235

List of Boxes

● ● ●

Box 7.1. Case studies of early/abrupt weaning of already vulnerable infants 163
Box 7.2. Considerations of wealth, status, and ethnic identity in childcare practices and access to emergency feeding programs 164

INTRODUCTION

● ● ●

A Biocultural Approach to Human Diet and Nutrition

T. Moffat and T. Prowse

Nothing would be more tiresome than eating and drinking if God had not made them a pleasure as well as a necessity.

—Voltaire (1769)

There are not many other topics we can think of that are more rooted in *both* the biological and social-cultural aspects of humankind. Our bodies need food and water to survive, but what we eat, how we prepare it, who we consume it with, and what we throw away are all influenced by our cultural environment. Throughout history human nutrition has been shaped by political-economic and cultural forces, and in turn food and nutrition can alter the course and direction of human societies. As Voltaire's quote so drolly indicates, eating and drinking are both biological necessities *and* pleasurable preoccupations. At the same time, the need to obtain food and render it in a form that is both nutritious and palatable can be difficult and tedious, depending on one's environmental, political, social, and economic situation. Patterns of food production and consumption have evolved with our species, and change throughout the human life course, but we recognize that biology and culture are inextricably linked when we look at human diet and nutrition, past and present.

What may seem a deceptively simple idea, however, is the challenge of conceptualizing, studying, and communicating the biocultural nature of human diet and nutrition. A survey of the fields that examine food, diet, and nutrition shows that they are splintered and for the most part isolated in their research endeavors. These range from nutrition sciences—largely biochemical in purview—that focus on the functional requirements of the human body and how our biological needs relate to what we eat, to the other end of the spectrum—studies of food and eating in social-cultural anthropology and sociology that focus almost exclusively on the structure and meaning of food use and values (e.g., Levi-Strauss 1970, 1973, 1978; Douglas 1972, 1984; for a review of some of this literature see Mintz and Dubois 2002). While these fields are crucial in the advancement of the study of human nutrition, they almost never converge, and therefore do not adequately address the biocultural aspects of diet and nutrition.

Anthropologists come closest in their attempts to consider the broad biocultural aspects of human diet and nutrition; here we briefly review some of the key examples from the literature that have been formative in the development of biocultural perspectives on human diet and nutrition. Pelto and colleagues (2000: 1) state that "nutritional anthropology is fundamentally concerned with understanding the interrelationships of biological and social forces in shaping human food use and the nutritional status of individuals and populations." They present an ecological model, originally conceived by Jerome et al. (1980)—including culture and environment, food, nutrients, nutritional status, and functional outcomes—to best capture the holistic nature of the field. Within the realm of culture and environment, there is a decidedly materialist bent in prioritizing political-economy as a major influence on the nutritional health outcomes of human populations (Goodman and Leatherman 1998). Excellent examples of this approach are: the edited volume by Cohen and Armelagos (1984) that examines the impact of agriculture on the diet and health of prehistoric societies; Leonard's (1991, 1992) investigations of contemporary Andean peoples' diet and nutritional status; Dufour's studies of bitter cassava production and consumption among Amazonian peoples (1995) and Dufour et al.'s (1997) description of food insecurity among Colombian women; and Pelto's (1987) work on dietary change among contemporary Mexicans.

Other areas of nutritional anthropology that have informed the field are: evolutionary studies of early hominid diets (see below), food systems and subsistence patterns such as studies of hunter-gatherers (Lee and Devore 1968) and optimal foraging (Boone 2002), and the evolution of agriculture (Armelagos and Harper 2005). Studies of "dietary delocalization" (Pelto and Pelto 1983) and the globalization of single-commodity foods such as sugar (Mintz 1985), as well as applied anthropological studies that focus on alleviating malnutrition (Pelto 2000), have also been influential.

Despite the works cited above being united in the field of anthropology, they are disparate, and it is challenging to describe a coherent field of "nutritional anthropology." Even our own subfield of biological anthropology is bifurcated in its division between studies of diet and nutrition in the past and present, as if somehow there are no connections. Thus, it is with a conscious effort that we have designed this book to show the breadth of the field related to the study of food in human societies on a chronological continuum, from broad-scale evolutionary time (the longue durée) down to individual time (the life course).

This edited volume is a collection of contributions by anthropologists studying aspects of diet and human nutrition through space and time using a biocultural approach. Many of the chapters employ a life-course perspective that recognizes the interrelationship between biological and social processes and the connections between different stages in the life span (Giele and Elder 1998; Kuh and Ben-Schlomo 2004). Chapter topics span prehistoric, historic, and contemporary societies and are situated in a variety of geographical regions, including Europe, North America, Africa, and Asia. Some of the chapters are reviews of important areas of research in the biocultural study of diet and nutrition; while others are original research articles that provide the reader with examples of an applied biocultural approach. Several of the chapters investigate diet and nutrition in skeletal samples using innovative biochemical techniques that are now standard practice in studies of diet in past populations, such as isotopic (chapters 4 and 8) and histological (chapter 8) analyses of bones and teeth; others rely on the markers of growth among both skeletal (chapters 2 and 8) and living (chapter 7) groups. Others make use of more traditional methods from anthropology such as participant observation, interviews, and focus groups (chapter 7) and the use of historical archival data (chapter 5). One chapter employs a methodology from nutritional sciences, the twenty-four-hour dietary recall, and embeds this analysis in a discussion of the biocultural factors that contribute to food choices (chapter 6). All of the chapters, however, are connected in their investigation of how the social and behavioral dynamics of human societies influence human food consumption and nutrition, as well as their functional outcomes.

Our aim is to open a dialogue and to encourage further conversations in publications and at conferences, to converge on the topic of diet and nutrition as a whole, rather than remain in our specialized subfields. In this volume we present themes and approaches to the biocultural study of diet and nutrition that intersect, and we now turn to a review of those themes.

Evolutionary Studies

The study of human nutrition within an evolutionary framework has been an enduring focus in biological anthropology, and researchers continue to investigate the impact of nutrition on human evolution. Methods of food acquisition and patterns of consumption are central components in models of hominid origins and evolution (Hockett and Haws 2003; Ungar et al. 2006). We are, in essence, what our ancestors ate (Zihlman 1976). Key stages in human evolution such as the origins of bipedalism (Lovejoy 1981; Hunt 1996), tool use (Lee and DeVore 1968), origins of social organization (Zihlman and Tanner 1978), and brain expansion (Aiello and Wheeler 1995) have been linked to the acquisition and consumption of food. More recently, researchers have focused on the "dietary capabilities" of the earliest hominids (Teaford and Ungar 2000: 13506). The chapters in this section exemplify the utility of integrating various lines of evidence for understanding the impact of diet and nutrition on our species within an evolutionary framework.

Leonard and colleagues (chapter 1) argue that the evolution of human brain size had important nutritional and metabolic consequences. They point to the paleontological evidence that indicates that the first major "pulse" of brain evolution in the hominid lineage occurred with the emergence of the genus *Homo* at 2.0–1.0 million years ago. They posit that humans can only support such a large, metabolically expensive brain with a high-quality, nutritionally dense diet. Concomitant with these changes were an increase in body size and an increase in foraging ecology that necessitated higher energy expenditure and the movement into larger ranges. They argue that an understanding of this profound evolutionary shift has implications for modern-day nutritional problems such as linear (height) growth stunting among children in developing countries and obesity in the contemporary global context.

Moving to a much later period in human evolution, Pfeiffer and Harrington (chapter 2) investigate the pattern and rate of childhood growth among Holocene (ca. 10,000 BP) foragers in South Africa. They use a combination of archaeological evidence, isotopic data, ethnographic analogy, and analyses of cranial and postcranial growth patterns to explore whether growth during this period was compromised due to dietary insufficiency. They demonstrate that although smaller in overall size than other human populations, these Stone Age children did not experience significant growth stunting. They hypothesize that the children had adequate nourishment and may have grown slowly with less magnitude because of a physical environment where being small in size was advantageous.

Finally, Sellen (chapter 3) links the evolutionary section to the following one on infant and young child feeding in a review of mammalian lactation and weaning patterns using a life history approach. Life history is the organization and temporal sequencing of major life events, such as age at first reproduction or death, that vary from species to species as an adaptive response to natural selection (See chapter 3 for a more detailed definition). As the only primate that weans juveniles before they can independently forage, this early and flexible weaning pattern in humans accounts for our high rate of fertility, due primarily to a shorter interbirth interval. A unique feature of human life history is transitional feeding (the addition of non–breast milk liquids and solids to breastfeeding), which he argues, based on current physiological knowledge and epidemiology, evolved among humans to begin at approximately six months after birth.

Breastfeeding and Beyond—Nutrition throughout the Life Course

Despite the evidence for an ideal evolved pattern of infant feeding among humans, as Sellen points out (chapter 3), humans are notoriously labile in their feeding practices, with patterns and differential timing of breastfeeding and weaning influenced by ecology, political-economy, and cultural norms and values. Anthropologists have conducted a number of empirical studies documenting variability in infant feeding practices and its effect on growth and development in past and present societies (See Dettwyler and Fishman 1992; VanEsterik 2002 for reviews). The study of breastfeeding and beyond to transitional foods and weaning is an important area for anthropologists and health professionals, as infants are at the most vulnerable stage in the life course. The influence of infant feeding on human fertility and mortality is of longstanding interest to biological anthropologists (Vizthum 1994; Ellison 1995; Katzenberg et al. 1996); however, there has been a growing demand in all areas of anthropology to examine the lives of infants and children in their own right (e.g., Panter-Brick 1998; Perry 2005; Lewis 2007).

It is challenging to study infant feeding in past human societies, particularly those without written texts or limited documentation, and anthropologists have developed a number of methodological innovations to uncover this information. For example, Dupras (chapter 4) explores infant feeding and weaning practices in the Dakhleh Oasis, Egypt (ca. first to fifth century CE) using stable isotope analysis of bones and teeth. Stable isotopes are frequently used to study breastfeeding and weaning patterns in past human populations, usually involving the analysis of nitrogen and carbon in bone samples. Dupras discusses cross-sectional data from subadult bone samples, which reveal that infants were breastfed until six months of age and not weaned until around three years of age. She then discusses the analysis of isotopes in tooth enamel and dentin from individuals in this skeletal sample, providing longitudinal data from the early-forming enamel and the later-forming dentin. These complementary lines of evidence are then integrated with Roman literary sources and archaeological evidence to reconstruct infant feeding and diet in this Roman period sample.

The other important aspect of infant feeding is the biocultural dimension of mothers' behaviors and how they are influenced by the societal context, including ideology and political-economy. Sawchuk and colleagues (chapter 5) examine secular trends in breastfeeding behaviors in the British territory of Gibraltar between 1955 and 1994. Their investigation integrates historical information with archival evidence from the local hospital's maternity registers. They examine breastfeeding trends of primaparous mothers in relation to sociocultural, political, and economic transitions, particularly in relation to governmental changes in border and labor force policies. Their chronological analysis shows clear shifts in the proportion of breastfeeding mothers during different periods of Gibraltar's history, and effectively demonstrates the impact of large-scale political-economic changes on human behavior, and ultimately on the diet and nutrition of infants.

Food Insecurity in the Developing World

There is a longstanding tradition of nutritional anthropologists focusing on the developing world, examining malnutrition, food insecurity, and policy. This began with colonial and postcolonial studies in the Third World and continues with research on diet and nutrition linked to malnutrition, food security, agriculture, and food policy (Turshen 1983; Huss-Ashmore and Johnston 1985; DeWalt 1998; Pelto and Pelto 1989).

It is now well accepted that human hunger is not a result of too little food production, but rather a question of access in combination with entitlement (Drèze and Sen 1991). Food insecurity and its impact on food access and nutrition in the developing world is a theme explored in the next two chapters. Anderson (1990: 1575–76) states, "Food insecurity exists whenever the availability of nutritionally adequate and safe food or the ability to acquire acceptable foods in socially acceptable ways is limited or uncertain." Both of these chapters demonstrate the importance of including "local voices" about food security in addition to quantitative analyses of food access.

Moffat and Finnis (chapter 6) compare dietary diversity among children (0–5 years) in periurban and rural Nepal. Dietary diversity data were collected for the two groups, and are analyzed within the context of the nutritional and economic consequences of dietary transitions associated with the phenomenon of rural-to-urban migration. In contrast to widely reported trends of increased obesity in modernizing urban environments in the developing world (Popkin and Gordon-Larsen 2004), the authors demonstrate that chronic *under*nutrition persists in urban environments and can be a consequence of dietary delocalization. The authors conclude by exploring the "moral economies" of food choices. They ask how do values and preferences affect food acquisition and ultimately the food that children consume?

Casiday and colleagues (chapter 7) present a case study from Niger, an impoverished nation in Central West Africa. The authors examine the effectiveness of local, regional, and international responses to the Sahelien food crisis in 2005. They consider the multiple factors that were responsible for the inadequate response to the food crisis by both the Niger government and international aid agencies. The authors then explore local reactions to the food crisis, focusing on parental perceptions of feeding and therapeutic food programs for children. This case study reveals the dramatic impact of political and economic factors on chronic food insecurity and emergency food crises.

Nutritional Factors in Growth and Disease

Diet and nutrition have functional outcomes that result in health and disease. Human health cannot be maintained without adequate macro- and micronutrients. There are a number of disease outcomes that are related to malnutrition including infectious disease (Waterlow and Tomkins 1992), chronic diseases such as cancer, cardiovascular disease, and inflammatory and autoimmune diseases (Simopoulos 2006). Conversely, an overabundance of energy-dense foods can result in disease outcomes associated with the state of obesity such as Type-2 diabetes (Barness et al. 2007). It is now widely recognized that early developmental stress can have a significant impact on later morbidity and mortality (Humphrey and King 2000; Kuh and Ben-Schlomo

2004; Kuzawa 2005). There is, however, a complex relationship between nutrition, human health, and disease throughout the life course, and other important variables such as growth, pregnancy and lactation, and the environment in addition to diet must be taken into account.

The first two chapters in this section use skeletal samples to reconstruct past lifeways to identify variables that influence health and disease of past populations. Prowse and colleagues (chapter 8) employ a multifaceted approach to explore diet, nutrition, and growth in a Roman period skeletal sample from Isola Sacra (first to third century CE), Italy. Their study integrates histological, stable isotope, and long bone growth data with historical evidence for infant and childhood feeding practices in Roman Italy. The combined histological and isotopic data indicate early diet-related stress, but the growth data reveal that these children did not suffer significant growth delay until late childhood. The authors conclude their analysis by examining the evidence in light of written records from the Roman period that address infant and childhood feeding practices.

Agarwal and Glencross (chapter 9) provide a comprehensive review of factors that contribute to bone maintenance and fragility throughout the life course. They consider both archaeological and contemporary studies to discuss the relationship between nutrition and bone mass, in particular those nutrients that are responsible for growth and maintenance of bone (e.g., protein, calcium, vitamin D). They address major life course events that can have an impact on bone mass, such as pregnancy and lactation, and explain how they can be detected in human skeletal remains. Finally, the authors address the relationship between nutrition and osteoporosis-related bone fractures, which has received considerable attention in studies of both past and present populations.

The final chapter of this section examines a uniquely contemporary phenomenon, the remarkable global increase in obesity in the late twentieth and twenty-first centuries, which is increasingly being problematized by anthropologists (Ulijaszek and Lofink 2006). Katzmarzyk (chapter 10) reviews the growing trend towards obesity in North American populations. The author provides a clear explanation of epidemiological definitions of "overweight" and "obese," and discusses temporal trends of obesity among children and adults in Canada, the United States, and Mexico. Katzmarzyk reviews the disease burden, including morbidity and mortality risks associated with obesity, and concludes by discussing the direct and indirect costs of obesity on North American health care systems.

In the concluding chapter, we link studies of the past and present and extend themes from the previous chapters to suggest new areas for the biocultural study of diet and nutrition. These themes are yet to appear or are emerging, but may hold promise for the future of this field.

Conclusion

The biocultural approach in nutritional anthropology is a holistic perspective that includes the use of evolutionary, ecological, cultural, and political-economic frameworks to investigate human nutrition in the social context. The chapters in this book span human history and geography; they point to both the diversity and shared human elements of our diets and how our ecologies, choices, and constraints affect our diets and nutrition as a species.

References

Aiello, L. C., and P. Wheeler. 1995. The expensive tissue hypothesis: The brain and the digestive system in human evolution. *Current Anthropology* 36: 199–221.

Armelagos, G. J., and K. N. Harper. 2005. Genomics at the origins of agriculture, part one. *Evolutionary Anthropology* 14: 68–77.

Anderson, S. A. 1990. Core indicators of nutritional state for difficult-to-sample populations. *Journal of Nutrition* 120 (11 suppl): 1559–1600.

Barness, L. A., J. M. Opitz, and E. Gilbert-Barness. 2007. Obesity: Genetic, molecular, and environmental aspects. *American Journal of Medical Genetics* 143A: 3016–34.

Boone, J. L. 2002. Subsistence strategies and early human population history: An evolutionary ecological perspective. *World Archaeology* 34 (1): 6–25.

Cohen, M. N., and G. J. Armelagos, eds. 1984. *Paleopathology at the Origins of Agriculture*. London: Academic Press.

DeWalt, B.R. 1998. The political ecology of population increase and malnutrition in Southern Honduras. In *Building a New Biocultural Synthesis. Political-Economic Perspectives on Human Biology,* ed. A. H. Goodman and T. H. Leatherman, 295–316. Ann Arbor: University of Michigan Press.

Dettwyler, K. A., and C. Fishman. 1992. Infant feeding practices and growth. *Annual Review of Anthropology* 21: 171–282.

Douglas, M. 1972. Deciphering a meal. *Deadalus, Journal of American Academy of Arts and Science* 101: 61–81.

———. 1984. *Food in the Social Order: Studies of Food and Festivities in Three American Communities*. New York: Russel Sage Foundation.

Drèze, J., and A. Sen. 1989. *Hunger and Public Action*. Oxford: Oxford University Press.

Dufour, D. L. 1995. A closer look at the nutritional implications of bitter cassava use. In *Indigenous Peoples and the Future of Amazonia: An Ecological Anthropology of an Endangered World,* ed. L. Sponsel, 149–65. Tucson: University of Arizona Press.

Dufour, D. L., L. K. Staten, J. C. Reina, and G. B. Spurr. 1997. Living on the edge: Dietary strategies of economically impoverished women in Cali, Colombia. *American Journal of Physical Anthropology* 102: 5–15.

Ellison, P. T. 1995. Breastfeeding, fertility, and maternal condition. In *Breastfeeding: Biocultural Perspectives,* ed. P. Stuart-Macadam and K. T. Dettwyler, 305–46. New York: Aldine de Gruyter.

Giele, J. Z., and G. H. Elder, Jr. 1998. *Methods of Life Course Research: Qualitative and Quantitative Approaches*. London: Sage Publications.

Goodman, A. H., and T. H. Leatherman, eds. 1998. *Building a New Biocultural Synthesis: Political-Economic Perspectives on Human Biology*. Ann Arbor: University of Michigan Press.

Hockett, B., and J. Haws. 2003. Nutritional ecology and diachronic trends in Paleolithic diet and health. *Evolutionary Anthropology* 12: 211–16.

Humphrey, L. T., and T. King. 2000. Childhood stress: A lifetime legacy. *Anthropologie* 37 (1): 33–49.

Hunt, K. D. 1996. The postural feeding hypothesis: An ecological model for the evolution of bipedalism. *South African Journal of Science* 92: 77–90.

Huss-Ashmore, R., and F. E. Johnston. 1985. Bioanthropological research in developing countries. *Annual Review of Anthropology* 14: 475–528.

Jerome, N. W., R. F. Kandel, and G. H. Pelto. 1980. An ecological approach to nutritional anthropology. In *Nutritional Anthropology: Contemporary Approaches to Diet and Culture*, ed. N. W. Jerome, R. F. Kandel and G. H. Pelto, 13–45. Pleasantville, NY: Redgrave Publishing Co.

Katzenberg, M. A., D. A. Herring, and S. R. Saunders. 1996. Weaning and infant mortality: Evaluating the skeletal evidence. *Yearbook of Physical Anthropology* 39: 177–99.

Kuh, D., and Y. Ben-Schlomo. 2004. *A Life Course Approach to Chronic Disease Epidemiology*. 2nd ed. Oxford: Oxford University Press.

Kuzawa, C. 2005. Fetal origins of developmental plasticity: Are fetal cues reliable predictors of future nutrition? *American Journal of Human Biology* 17:5–21.

Lee, R., and I. Devore, eds. 1968. *Man the Hunter*. Chicago: Aldine.

Leonard, W. R. 1991. Household-level strategies for protecting children from seasonal food scarcity. *Social Science and Medicine* 33(10): 1127–1133.

———. 1992. Variability in adaptive responses to dietary change among Andean farmers. In *Health and Lifestyle Change. MASCA Research Papers, Volume 9*, ed. R. Huss-Ashmore, J. Schall, and M. Hediger, 71–82. Philadelphia: MASCA The University Museum of Archeology and Anthropology, University of Pennsylvania.

Lewis M. E. 2007. *The Bioarchaeology of Children: Perspectives from Biological and Forensic Anthropology*. Cambridge: Cambridge University Press.

Levi-Strauss, C. 1970. *The Raw and the Cooked*. London: Cape.

———. 1973. *From Honey to Ashes*. London: Cape.

———. 1978. *The Origin of Table Manners*. London: Cape.

Lovejoy, C. O. 1981. The origins of man. *Science* 211: 341–348.

Mintz, S. W. 1985. *Sweetness and Power: The Place of Sugar in Modern History*. New York: Penguin.

Mintz, S. W., and C. M. Dubois. 2002. The anthropology of food and eating. *Annual Review of Anthropology* 31: 99–119.

Panter-Brick, C. 1998. *Biosocial Perspectives on Children*. Cambridge: Cambridge University Press.

Pelto, G. H. 1987. Social class and diet in contemporary Mexico. In *Food and Evolution: Toward a Theory of Human Food Habits*, ed. M. Harris and E. B. Ross, 517–40. Philadelphia: Temple University.

———. 2000. Continuities and new challenges in applied nutritional anthropology. *Nutritional Anthropology* 23 (2): 16–22.

Pelto, G. H., A. H. Goodman, and D. L. Dufour. 2000. The biocultural perspective in nutritional anthropology. In *Nutritional Anthropology. Biocultural Perspectives on Food and Nutrition,* ed. G. H. Pelto, A. H. Goodman, and D. L. Dufour, 1–9. Mountain View, CA: Mayfield Publishing.

Pelto, G. H., and P. J. Pelto 1983. Diet and delocalization: Dietary changes since 1750. *Journal of Interdisciplinary History* 14 (2): 507–28.

———. 1989. Small but healthy? An anthropological perspective. *Human Organization* 48 (1): 11–15.

Perry M. A. 2005. Redefining childhood through bioarchaeology: Toward an archaeological and biological understanding of children in antiquity. *Archeological Papers of the American Anthropological Association* 15 (1): 89–111.

Popkin, B. M., and P. Gordon-Larsen. 2004. The nutrition transition: Worldwide obesity dynamics and their determinants. *International Journal of Obesity* 28: S2–S9.

Simopoulos, A. P. 2006. Evolutionary aspects of diet, omega-6/omega-3 ratio and genetic variation: Nutritional implications for chronic diseases. *Biomedicine & Pharmacotherapy* 60: 502–7.

Teaford, M. 2002. Paleontological evidence for diets of Africa Plio-Pleistocene hominins with special reference to early *Homo.* In *Human Diet: Its Origin and Evolution,* ed. P. Ungar and M. Teaford, 143–66. Westport, CT: Bergin & Garvey.

Teaford, M. F., and P. S. Ungar. 2000. Diet and the evolution of the earliest human ancestors. *Proceedings of the National Academy of Sciences* 97: 13506–11.

Turshen, M. 1983. Study of women, food, and health in Africa. In *Third World Medicine and Social Change: A Reader in Social Science and Medicine,* ed. J. H. Morgan, 241–51. Lanham, MD: University Press of America.

Ulijaszek, S. J., and H. Lofink. 2006. Obesity in biocultural perspective. *Annual Review of Anthropology* 35: 337–60.

Ungar, P. S., F. E. Grine, and M. F. Teaford. 2006. Diet in early *Homo*: A review of the evidence and a new model of adaptive versatility. *Annual Review of Anthropology* 35: 209–28.

VanEsterik, P. 2002. Contemporary trends in infant feeding research. *Annual Review of Anthropology* 31: 257–78.

Verhaegen, M., and P. F. Puech. 2000. Hominid diet and lifestyle reconsidered: Paleo-enviromental and comparative data. *Human Evolution* 15 (3–4): 175–86.

Vizthum, V. J. 1994. Comparative study of breastfeeding structure and its relation to human reproductive ecology. *Yearbook of Physical Anthropology* 37: 307–49.

Voltaire. 1769. *Les Adorateurs, Oeuvres Complètes de Voltaire, vol. 28,* ed. Louis Moland (1877–1885). Paris: Garnier.

Waterlow, J. C., and A. M. Tomkins. 1992. Nutrition and infection. In *Protein-energy Malnutrition,* ed. J. C. Waterlow, 290–324. London: Edward Arnold.

Zihlman, A. 1976. Sexual dimorphism and its behavioral implications in early hominids. *IX Congrès—Colloque VI: Les Plus Anciens Hominides.* P.V. Tobias, and Y. Coppens. CNRS, Paris: 268–293.

Zihlman, A. and N. Tanner 1978. Gathering and the hominid adaptation. *Female Hierarchies.* L. Tiger, and H. Fowler. Chicago, Beresford Books, 163–194.

Evolutionary Perspectives on Nutrition

• 1 •
What Did Humans Evolve to Eat?
Metabolic Implications of Major Trends in Hominid Evolution

W. R. Leonard, M. L. Robertson, and J. J. Snodgrass

Introduction

Over the last twenty years, the evolution of human nutritional requirements has received ever-greater attention among both anthropologists and nutritional scientists (Crawford 1992; Eaton and Konner 1985; Garn and Leonard 1989; Leonard and Robertson 1992, 1994; Aiello and Wheeler 1995; Cordain et al. 2005; Ungar 2007). Research in nutritional anthropology has demonstrated that many of the key features that distinguish humans from other primates have important implications for our distinctive nutritional needs (Leonard 2002; Leonard and Robertson 1997b; Aiello and Wheeler 1995). In addition, our colleagues in the nutritional sciences are coming to realize that an evolutionary perspective is useful for understanding the origins of and potential solutions to the growing problems of obesity and associated metabolic disorders (e.g., Cordain et al. 2005; Eaton 2006; O'Dea 1991).

Yet, despite this growing consensus that an evolutionary approach has an important place in the study of human nutrition, we find that many constructions of the "natural" human diet are remarkably narrow (e.g., Audette and Gilchrist. 2000; Crawford and Marsh 1995; Cunanne 2005; Eaton, Shostack, and Konner 1988). We believe that many of these "paleodiets" are based on a misreading of both human evolutionary history and comparative human biology. Humans did not evolve to subsist on a single Paleolithic diet. To the contrary, one of the hallmarks of our evolutionary success has been our ability to find or create a meal in any environment. Compared to other primates, humans have diets of much higher quality—that is, more dense in calories and nutrients. Indeed, many of the major changes in human evolutionary history have been about increasing the quality of our diets or increasing the efficiency with which we extract energy and nutrients from our environments.

This chapter specifically considers the nutritional implications of one of the most profound transition periods in human evolution—the emergence of the first members of the genus *Homo*. This phase of human evolution—between ~2.0 and 1.5 million years ago—was associated with major changes in brain size, body size, and foraging and ranging behavior.

To establish the context for interpreting the fossil evidence, we begin by considering the energetic and nutritional correlates of variation in brain and body size among living primates. We then turn to an examination of the human fossil record to consider when and under what conditions in our evolutionary past key changes in brain size, body size, diet, and foraging behavior likely took place. Finally, we explore the implications of our distinctive metabolic requirements for understanding and confronting the nutritional problems of our modern world. We will specifically consider (1) the problem of early childhood growth stunting among populations of the developing world, and (2) the growing problem of obesity in the United States and other industrialized nations.

Comparative Nutrition and Metabolism

From a nutritional perspective, what is extraordinary about our large human brains is their high metabolic costs. Brain tissue has very high energy demands per unit weight, roughly 16 times greater than those of muscle tissue (12 kcal/kg/min versus 0.75 kcal/kg/min; Holliday 1986; Kety 1957). On average about 400 kcal/day are spent on brain metabolism by an adult human. Yet, despite the fact that humans have much larger brains per body weight than other primates or terrestrial mammals, the resting energy demands for the human body are no more than for any other mammal of the same size (Leonard and Robertson 1994).

Figure 1.1a shows the relationship between Resting Metabolic Rate (RMR; kcal/day) and body mass in kilograms (kg) for humans and nonhuman primates, and other mammals. It is clear that humans, as well as other primate species, conform to the general mammalian scaling relationship between RMR and body weight, the "Kleiber Relationship" (Kleiber 1961). The Kleiber Relationship shows that metabolic rates in mammals of vastly different sizes increase as a function of body weight raised to the 3/4th power. Thus for a mammal of a given body mass, we can predict their resting energy needs as:

RMR (kcal/day) = $70(Wt^{0.75})$

On average, adult humans have RMRs that fall within 3–4 percent of the value predicted for other primates and other mammals. The implication of this is that humans allocate a much larger share of our daily energy budget for brain metabolism than other species.

The disproportionately higher energy costs of our large brains are evident in the scaling relationship between brain weight (grams) and RMR for humans, thirty-five other primate species, and twenty-two nonprimate mammalian species (Fig. 1.1b). The solid line denotes the best-fit regression for nonhuman primate species, and the dashed line denotes the best-fit regression for the nonprimate mammals. The data point for humans is denoted with a star.

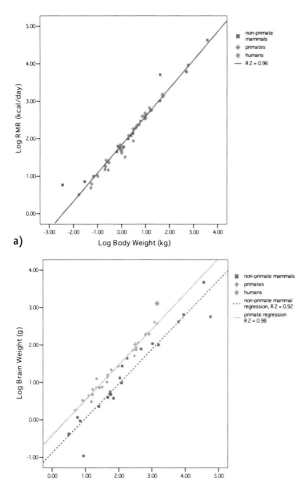

Figure 1.1. a) Log-Log plot of resting metabolic rate (RMR; kcal/day) versus body weight (kg) for 51 species of terrestrial mammals (20 non-primate mammals, 30 primates, and humans). Humans conform to the general mammalian scaling relationship, as described by Kleiber (1961); b) Log-Log plot of brain weight (BW;g) versus RMR (kcal/day) for humans, 35 other primate species, and 22 non-primate mammalian species. The primate regression line is systematically and significantly elevated above the non-primate mammal regression. For a given RMR, primates have brain sizes that are three times those of other mammals, and humans have brains that are three times those of other primates.

As a group, primates have brains that are approximately 3 times the size of other mammals (relative to body size). Human brain sizes, in turn, are some 2.5 to 3 times those of other primates (Martin 1989). In caloric terms, this means that brain metabolism accounts for ~20–25 percent of RMR in an adult human body, as compared to about 8–10 percent in other primate species, and roughly 3–5 percent for nonprimate mammals (Leonard et al. 2003).

The large allocation of our energy budget to brain metabolism raises the question of how humans are nutritionally able to accommodate the metabolic demands of our large brains. It appears that humans consume diets that are more dense in energy and nutrients than other primates of similar size.

Across all primates, diet quality is inversely related to body size. That is, small primates (e.g., the pygmy marmoset) consume diets that are rich in energy and nutrients, whereas large-bodied primates (e.g., the gorilla) consume large amounts of low-quality foods (Richard 1985). These feeding strategies are shaped by between-species variation in metabolic rates, specifically the Kleiber Relationship, mentioned previously.

Small-bodied primates have low total energy needs but very high energy demands per unit mass (i.e., kcal/kg/day). Consequently, they meet their dietary needs by consuming foods that are limited in abundance but high in quality (insects, saps, gums). Large primates have high total energy need, but low mass-specific costs. Hence they are large-volume feeders, eating foods that are widely available, but of low nutritional density (leaves, bark, and low-quality plant foods).

Humans, however, have substantially higher-quality diets than expected for a primate of our size. Figure 1.2 shows the association between dietary quality and body weight in living primates, including modern human foragers. The diet quality (DQ) index is derived from the work of Sailer et al. (1985) and reflects the relative proportions (percentage by volume) of (1) structural plant parts (*s*; e.g., leaves, stems, bark), (2) reproductive plant parts (*r*; e.g., fruits, flowers), and (3) animal foods (*a*; including invertebrates):

DQ index = s + 2(r) + 3.5 (a)

The index ranges from a minimum of 100 (a diet of all leaves and/or structural plant parts) to 350 (a diet of all animal material).

There is a strong inverse relationship between DQ and body mass across primates; however, note that the diets of modern human foragers fall substantially above the regression line in Figure 1.2a. Indeed, the staple foods for all human societies are much more nutritionally dense than those of other large-bodied primates. Although there is considerable variation in the diets of modern human foraging groups, recent studies have shown that modern human foragers typically derive over half of their dietary energy intake from animal foods (Cordain et al. 2000). In comparison, modern great apes obtain much of their diet from low-quality plant foods. Gorillas derive over 80 percent of their diet from fibrous foods such as leaves and bark (Richard

1985). Even among common chimpanzees (*Pan troglodytes*), only about 5–10 percent of their calories are derived from vertebrate animal foods (Teleki 1981; Stanford 1996). This "higher-quality" diet means that we need to eat a lower volume of food to get the energy and nutrients we require.

The link between brain size and dietary quality is evident in Figure 1.2b, which shows relative brain size versus relative dietary quality for the thirty-three different primate species for which we have metabolic, brain size, and dietary data. Relative brain size for each species is measured as the standardized residual (z-score) from the primate brain versus body mass regression, and relative DQ is measured as the residual from the DQ versus body mass regression. There is a strong positive relationship (r = 0.63; P < 0.001) between the amount of energy allocated to the brain and the caloric and nutrient density of the diet. Across all primates, larger brains require higher-quality diets. Humans fall at the positive extremes for both parameters, having the largest relative brain size and the highest quality diet.

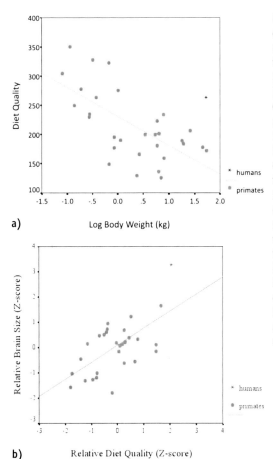

Figure 1.2. a) Plot of diet quality (DQ) versus log-body mass for 33 primate species. DQ is inversely related to body mass (r = −0.59 [total sample]; −0.68 [non-human primates only]; P < 0.001), indicating that smaller primates consume relatively higher quality diets. Humans have systematically higher quality diets than predicted for their size; b) Plot of relative brain size versus relative diet quality for 31 primate species (including humans). Primates with higher quality diets for their size have relatively larger brain size (r = 0.63; P < 0.001). Humans represent the positive extremes for both measures, having large brain:body size and a substantially higher quality diet than expected for their size.

Thus, the high cost of the large, metabolically expensive human brain is partially offset by the consumption of an energy- and nutrient-rich diet. This relationship implies that the evolution of larger hominid brains would have necessitated the adoption of a sufficiently high-quality diet (including meat and energy-rich fruits) to support the increased metabolic demands of greater encephalization.

Evolutionary Trends in Diet, Brain Size, and Body Size

When we look at the human fossil record, we find that the first major burst of evolutionary change in hominid brain size occurs at about 2.0 to 1.7 million years ago, associated with the emergence and evolution of early members of the genus *Homo* (see Table 1.1). Prior to this, our earlier hominid ancestors, the australopithecines, showed only modest brain size evolution from an average of 400 to 510 cm^3 over a 2-million-year span from 4 to 2 million years ago. With the evolution of the genus *Homo* there is rapid change, with brain sizes of, on average, ~600 cm^3 in *Homo habilis* (at 2.4–1.6 mya) and 800–900 cm^3 in early members of *Homo erectus* (at 1.8–1.5 mya). Furthermore, while the relative brain size of *Homo erectus* did not reach the size of modern humans, it is outside of the range seen among other living primate species.

Table 1.1. Geological ages (millions of years ago), brain size (cm^3), estimated male and female body weights (kg), and estimated home range sizes (hectares) for selected fossil hominid species.

Species	Geological age (mya)	Brain size (cm^3)	Body Weight Male (kg)	Body Weight Female (kg)	Home Range (ha)
A. afarensis	3.9–3.0	438	45	29	40
A. africanus	3.0–2.4	452	41	30	38
A. boisei	2.3–1.4	521	49	34	47
A. robustus	1.9–1.4	530	40	32	38
Homo habilis (sensu strictu)	1.9–1.6	612	37	32	226
H. erectus (early)	1.8–1.5	863	66	54	480
H. erectus (late)	0.5–0.3	980	60	55	452
H. sapiens	0.4–0.0	1350	58	49	410

Data for brain size and body weights are from McHenry and Coffing (2000), except for *Homo erectus*. Early *H. erectus* brain size is the average of African specimens as presented in McHenry (1994b), Indonesian specimens from Antón and Swisher (2001) and Georgian specimens from Gabunia et al. (2000, 2001). Brain size and body weight data for late *H. erectus* are from McHenry (1994a). Estimates of home range size are from Antón, Leonard and Robertson (2002).

The evolution of *H. erectus* in Africa is widely viewed as a major adaptive shift in human evolution (Antón 2003; Antón, Leonard, and Robertson 2002; Wolpoff 1999). Indeed, what is remarkable about the emergence of *H. erectus* in East Africa at 1.8 million years is that we find (a) marked increases in both brain and body size, and (b) the evolution of human-like body proportions at the same time that we see (c) major reductions of posterior tooth size and craniofacial robusticity (McHenry 1992, 1994a, 1994b; McHenry and Coffing 2000; Ruff, Trinkaus, and Holliday 1997). These trends clearly suggest major energetic and dietary shifts: (a) the large body sizes necessitating greater daily energy needs; (b) bigger brains suggesting the need for a higher-quality diet; and (c) the craniofacial changes suggesting that they were consuming a different mix of foods than their australopithecine ancestors.

The ultimate driving factors responsible for the rapid evolution of brain size, body size and craniodental anatomy at this stage of human evolution appear to have been major environmental changes that promoted shifts in diet and foraging behavior. The environment in East Africa at the Plio-Pleistocene boundary (2.0–1.8 mya) was becoming much drier, resulting in declines in forested areas and an expansion of open woodlands and grasslands (Bobe and Behrensmeyer 2002; deMenocal 2004; Reed 1997; Vrba 1995; Wynn 2004). Such changes in the African landscape likely made animal foods an increasingly attractive resource for our hominid ancestors (Behrensmeyer et al. 1997; Harris and Capaldo 1993; Plummer 2004).

This can be seen by looking at the differences in ecological productivity between modern-day woodland and savanna ecosystems of the tropics. Despite the fact that tropical savanna environments produce only about half as much plant energy per year as tropical woodlands (4050 vs. 7200 kcal/m^2/year), the abundance of herbivores (secondary productivity) is almost three times greater than in the savanna (10.1 vs. 3.6 kcal/m^2/year) (Leonard and Robertson 1997a). Consequently, the expansion of the savanna in Plio-Pleistocene Africa would have limited the amount and variety of edible plant foods (to things like tubers, etc.) for hominids, but also resulted in an increase in the relative abundance of grazing mammals such as antelope and gazelle. These changes in the relative abundance of different food resources offered an opportunity for hominids with sufficient capability to exploit the animal resources. The archeological record suggests that this is what occurred with *Homo erectus*—the development of the first rudimentary hunting and gathering economy in which (1) game animals became a significant part of the diet and (2) food resources were shared within foraging groups.

The other major evolutionary event seen with early *Homo erectus*—the rapid initial spread of hominids from Africa to other parts of the Old World—appears to be linked to changes in ecology and the associated changes in brain size, body size, and foraging behavior. In living species, we know that an important correlate of dispersal distance is territorial needs—day range (the distance [km] that an animal travels on a typical day) and home range sizes (the total area [hectares] utilized by an animal population). Species with relatively larger territories for their size have greater potential for more rapid dispersion. Additionally, we know that human foragers are distinct from other primate species in having very large territorial needs for their size (Antón, Leonard, and Robertson 2002; Leonard and Robertson 1997b).

Comparative studies on territorial needs and ranging behavior have shown that strongest predictors of variation in home range (HR) size are: (a) body mass (kg), (b) diet (with carnivores having much larger HR sizes than herbivores), and (c) ecosystem structure, with species living in more-open, less-productive habitats having larger territorial needs (Harestad and Bunnell 1979). With the emergence of *H. erectus* we see changes in all three of these parameters that would have promoted increased territorial requirements.

To model how changes in body size and diet may have influenced HR sizes in our hominid ancestors, we compiled data on body size, HR area, and DQ for forty-seven nonhuman primate species and six human hunting and gathering groups (data presented in Antón, Leonard, and Robertson 2002). HR size is strongly associated with body mass; however, human foragers have substantially larger HRs than other primates of their size.

Using a multiple regression approach, we explored the joint influences of weight and DQ on variation HR size. When diet was included along with body mass in the prediction, the model explained 77 percent of the variation in HR size (see Antón, Leonard, and Robertson 2002). In light of the high predictive power of the model, it is useful for exploring changes in the evolutionary past, since the model effectively applies to both human foragers and nonhuman primates.

Using the model derived from living species, we estimated evolutionary changes in HR size among prehistoric hominid species by: (a) using mid-sex estimates of body weight derived from McHenry and Coffing (2000), and (b) assuming a modest increase in dietary quality between the australopithecines and early *Homo*. Specifically, we assumed australopithecines to have DQ equal to the average for modern ape species, whereas early members of the genus *Homo* had DQ equal to the minimum value for contemporary foragers.

With the above-noted changes in body size and modest improvement in DQ with the evolution of the genus *Homo*, we find dramatic increases in HR sizes with the evolution of *H. erectus*. Estimated HR size for *H. erectus* is ~450 hectares, about 8 to 10 times that of their australopithecine ancestors. Thus the changes in body mass and dietary quality with origins of *H. erectus* would have dramatically influenced territorial needs and dispersal capability (see Table 1.1 for estimated home range sizes for selected fossil hominid species).

Thus we find several major, interrelated changes with the emergence of *Homo erectus*. With the expansion of the African grassland, there were declining levels of primary productivity and changes in resource distribution (i.e., more grazing animals and calories "on the hoof"). These changes appear to have promoted shifts in foraging behavior and dietary quality that helped to provide the energetic/nutrition fuel to support the rapid evolution of both brain size and body size. In addition, the lower ecological productivity, dietary change, and increased body size all would have contributed to greater HR needs and dispersal potential.

This adaptive package that we see with *H. erectus* highlights the evolution of key nutritional characteristics that are distinctly human: (a) evolution of our large brains, requiring a higher-quality, nutritionally dense diet, and (b) increased body size and the adoption of a foraging strategy that necessitated movement of large ranges that required high levels of daily energy expenditure.

Modern Human Nutritional Problems

Since the emergence of the genus *Homo* and the initial spread of hominids out of Africa, humans have successfully colonized almost every major ecosystem on the planet. Our ancestors' ability to exploit diverse environments was, in large measure, dependent upon developing strategies and technologies for increasing energy returns from subsistence activities, and raising the nutritional quality of staple food items. During the course of more recent human evolution, these strategies have included all of the following: (a) technological and foraging changes, (b) cooking, (c) development of agricultural and pastoral subsistence regimes, and (d) development of novel food processing/preparation techniques still seen today (e.g., alkali processing of maize in the Americas, potato processing/preservation in the Andes, processing of bitter manioc, processing of soy beans in Asia) (see Leonard 2000 for additional discussion). Today, we find that humans are able to subsist and thrive on a remarkable diversity of diets, ranging from those of arctic populations consisting almost entirely of animal material to those of many small-scale farming societies, subsisting almost exclusively on plant foods.

In the face of this enormous diversity, the features that are common across all these human groups are: (a) the ability to produce a nutritionally dense diet to support the demands of our large brains, and (b) the ability to obtain sufficient total energy to support our relatively large bodies and activity levels. Over our evolutionary history, we have been quite successful developing strategies for meeting our nutritional needs. However, even today many important health problems reflect the challenges that exist in accommodating our distinctive nutritional biology. Here we will explore two of these problems: early childhood growth stunting, and obesity and its associated metabolic disorders.

Linear Growth Stunting

The problem of growth stunting is, in part, driven by the extraordinarily high costs of brain metabolism in very young children. Whereas brain metabolism accounts for 20–25 percent of resting needs in adults, in an infant of < 10 kg, it is upwards of 60 percent (Holliday 1986)! Table 1.2 shows changes in the percent of RMR allocated

Table 1.2. Body weight (kg), brain weight (g), percent body fat (%), resting metabolic rate (RMR; kcal/day), and percent of RMR allocated to brain metabolism (BrMet, %) for humans from birth to adulthood.

Age	Body weight (kg)	Brain Weight (g)	Body fat (%)	RMR (kcal/day)	BrMet (%)
Newborn	3.5	475	16	161	87
3 months	5.5	650	22	300	64
18 months	11.0	1045	25	590	53
5 years	19.0	1235	15	830	44
10 years	31.0	1350	15	1160	34
Adult male	70.0	1400	11	1800	23
Adult female	50.0	1360	20	1480	27

All data are from Holliday (1986), except for percent body fat data for children 18 months and younger, which are from Dewey et al. (1993).

to the brain over the course of human growth and development. These enormously high energy demands for infants reflect both their high brain to body weight ratios and their rapid rates of brain growth.

To accommodate the extraordinary energy demands of the developing infant brain, human infants are born with an ample supply of body fat (Kuzawa 1998; Leonard et al. 2003). At ~15–16 percent body fat, human infants have the highest body fat levels of any mammalian species (Dewey et al. 1993). Further, human infants continue to gain body fat during their early postnatal life. During the first year, healthy infants typically increase in fatness from about 16 percent to about 26 percent (see Table 1.2).

Research on children of the developing world suggests that chronic, mild to moderate undernutrition has a relatively small impact on a child's fatness—that is, it appears to be preserved in the face of nutritional stress. Instead of taking away the fat reserves, nutritional needs appear to be down-regulated by substantially reducing rates of growth in height/length—producing the common problem of infant/childhood growth stunting or growth failure that is ubiquitous among impoverished populations of the developing world.

Figure 1.3 shows an example of this process based on growth data collected from the Tsimane' farmers and foragers of lowland Bolivia (from Foster et al. 2005). Note that stature early in life closely approximates the US median, but by age three to four years it has dropped below the 5th centile, where it will track for the rest of life. In contrast, body fatness (as measured by the sum of the triceps and subscapular skinfolds) compares more favorably to US norms, tracking between the 15th and 50th

US centiles. The problem of early childhood growth failure is the product of both increased infectious-disease loads and reduced dietary quality.

International health research has consistently shown that higher dietary quality is the strongest nutritional predictor of improved growth in young children of the developing world. For example, work with agricultural populations of highland and coastal Ecuador (Leonard et al. 2000; Berti, Leonard, and Berti 1998) and long-term studies carried out in Mexico (Allen 1994; Allen et al. 1992) and Guatemala (Martorell and Habicht 1986; Habicht, Martorell, and Rivera 1995) have all found that percent of dietary energy derived from animal sources was the single strongest predictor of growth velocities for children under the age of three years. In our research in Ecuador, we specifically found that home production and consumption of eggs

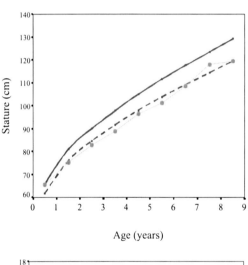

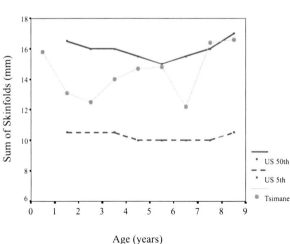

Figure 1.3. Patterns of physical growth in stature (cm) and body fatness (as sum of triceps and subscapular skinfolds, mm) in girls of Tsimane' of lowland Bolivia. Growth of Tsimane' girls is characterized by marked linear growth stunting, whereas body fatness compares more favorably to US norms. Data from Foster et al. (2005)

was strongly correlated with improved DQ and growth in young children (Leonard et al. 2000).

These findings highlight the distinctive nutritional constraints faced by human infants and toddlers—very high nutritional demands (driven by brain metabolism and the energy needs for rapid body growth) with a stomach of limited volume, placing a premium on foods of high nutritional density and digestibility.

Obesity and Chronic Metabolic Disorders

In contrast to the problems of undernutrition that continue to plague young children of the developing world, the United States and other industrialized nations are now experiencing unprecedented increases in obesity rates. In the United States, rates of obesity have increased dramatically over the last forty years. As shown in Table 1.3, US national data (NHANES data) show that since the early 1960s the prevalence of obesity (BMI ≥ 30.0 kg/m^2) has almost tripled in men and more than doubled in women. As of 2004, approximately a third of all Americans were classified as obese (31.1% of men; 33.2% of women). The most recent data suggest that obesity rates are continuing to climb in men, but may be leveling off in women (Flegal et al. 2002; Ogden et al. 2006). (See Katzmarzyk in this volume for a North American overview).

Two of the factors most often cited for the dramatic increases in obesity rates in the United States are the ever-greater availability of inexpensive high calorie/high fat food items, and expansion of "portion sizes" (i.e., the "Supersize Me" phenomenon).

Yet, while these changes in consumption patterns are certainly contributing to the problem, they are not nearly the entire story. In fact, national data show that there have been only modest increases in energy consumption over the last thirty years. This point is evident in Table 1.4, which shows mean daily calorie intakes for adult men and women between 1971 and 2000, derived from the US NHANES data (Briefel and Johnson 2004). Over this period—when body weights and obesity rates were dramatically increasing—daily energy intakes increased by ~200 kcal in men (8 percent increase) and ~300 kcal in women (20 percent increase). In addition, the percent of calories from fat actually declined during this span from 36 percent to 33 percent.

Table 1.3. Percentage of US Adults (20 years and older) who are obese (BMI ≥ 30 kg/m^2), 1960 to 2004.

Sample	Percent Obese (BMI > 30)					
	1960–1962	1971–1974	1976–1980	1988–1994	1999–2000	2003–2004
Men	10.7	12.1	12.7	20.6	27.5	31.1
Women	15.8	16.6	17.0	25.9	33.4	33.2

Sources: Flegal et al. (2002), Ogden et al. (2006)

Table 1.4. Dietary energy (kcal/day) and macronutrient (percent of dietary energy) intakes among US adults (20 years and older), 1971 to 2000.

Dietary Parameter	1971–1974	1976–1980	1988–1994	1999–2000
Energy (kcal/d)				
Males	2450	2439	2666	2618
Females	1542	1522	1798	1877
Macronutrients (% energy)				
Fat	36	36	34	33
Protein	15	15	14	14
Carbohydrates	49	49	52	53

Source: Briefel and Johnson (2004)

These data clearly suggest that the obesity epidemic cannot be understood solely by looking at the intakes; rather we must also consider energy expenditure and activity patterns. Unfortunately, because of the difficulty in measuring daily energy expenditure, we do not have national-level data on changes in expenditure and activity levels. However, a look at some comparative data on energy expenditure across populations with different lifestyles can give us a handle on how urbanization of lifestyles influences daily energy demands.

The most common approach to quantifying differences in "activity level" (metabolic intensity) associated with different lifestyles is simply by expressing Total Daily Energy Expenditure (TDEE) as a ratio relative to RMR. This ratio of TDEE to RMR is known as the Physical Activity Level (PAL) index. Based on comparative research conducted on human populations around the world, the WHO has established a range of PALs associated with different occupational workloads among adults (FAO/WHO/UNU 1985; FAO/WHO/UNU 2004). The PAL associated with minimal daily activities (simply dressing, washing, and eating) is 1.40 for both men and women. Sedentary lifestyles (e.g., office work) require PALs of 1.55 for men and 1.56 for women. At high workloads, the sex differences in PALs tend to be greater. Moderate workloads commensurate with PALs of 1.78 for men and 1.64 for women, whereas heavy occupational workloads (e.g., manual laborers, subsistence farmers during harvest periods) require PALs of 2.10 and 1.82 for men and women, respectively.

Human populations show considerable variation in levels of daily energy expenditure. Several recent comparative analyses indicate that daily energy expenditure in human groups typically ranges from 1.2 to 5.0 × RMR (i.e., PAL = 1.2–5.0) (Black et al. 1996; Institute of Medicine 2002). The lowest levels of physical activity, PALs of 1.20 to 1.25, are observed among hospitalized and nonambulatory populations.

The highest levels of physical activity (PALs of 2.5 to 5.0) have been observed among elite athletes and soldiers in combat training.

Table 1.5 presents data on body weight (kg), TDEE (kcal/day), RMR (kcal/day), and PALs of adult men and women from selected human groups. Values for individuals of the industrialized world are based on a sample of 258 men and 259 women aged twenty years or older compiled by the Institute of Medicine for their most recent references on dietary energy intakes (Institute of Medicine 2002). These data were derived from twelve studies of human energy expenditure using the doubly labeled water method, generally accepted as the most accurate technique for assessing TDEE in free living humans. The data for human subsistence-level populations (i.e., foragers, pastoralist animal herders, and agriculturalists) were derived from anthropological studies conducted over the last forty years.

Among subsistence-level populations, the average daily energy expenditure is about 3000 kcal/day for men and about 2300 kcal/day for women. The energy demands of life in the industrialized world are more modest. Men of the Western world sample are, on average, 12 kg (26.5 pounds) heavier than their counterparts from the subsistence populations, and yet have daily caloric needs that are 150 to 200 kcal less. The pattern is similar, although somewhat less dramatic, for women. Those of the industrialized world are 7 kg (15.5 pounds) heavier and have daily energy demands that are about 60 kcal less than those from food-producing societies.

This means that adults living a "modern" lifestyle in the industrialized world have lower physical activity levels than those living more "traditional" lives. Among men, PALs in the industrialized societies average 1.73, significantly less than the average of 1.98 among the subsistence-level groups ($P < 0.01$). Thus, the difference in the predicted TDEEs at a body weight of 70.1 kg (the average for the industrialized sample) is over 900 kcal (3784 vs. 2874 kcal/day) (see Figure 1.4). Similarly, PAL values among women average 1.72 in the industrialized world and 1.82 among the subsistence-level societies. The difference in predicted energy demands at the average body weight of the industrialized world sample (58.6 kg) is about 330 kcal (2565 vs. 2234 kcal/day). These large differences in daily energy demands underscore how the substantial reductions in intense physical activities have dramatically lowered the metabolic costs of survival in the modern world. The more modest declines in TDEE observed in women is partly a product of their smaller body size and partly a consequence of the large average sex-differences in TDEE and PAL that characterize traditional, subsistence-level societies. In subsistence-level societies, the most metabolically demanding activities (e.g., hunting, plowing fields) are generally performed by men. Consequently, the transition from a "traditional" to a more "modern" way of life is often associated with greater reductions in physical energy demands in men.

In sum, the daily energy demands in modern subsistence-level societies are considerably greater than those observed in the industrialized world. These comparisons suggest that the transition from a subsistence to a sedentary, modern lifestyle is associated with a 15–30 percent reduction in one's maintenance energy needs.

Table 1.5. Comparison of body weight (kg), total daily energy expenditure (TDEE; kcal/day), resting metabolic rate (RMR; kcal/day), and Physical Activity Level (PAL) of subsistence-level and industrial societies.

Group	Sex	Weight (kg)	TDEE (kcal/day)	RMR (kcal/day)	PAL	References
Hunter-gatherers:						
!Kung (Botswana)	M	46.0	2319	1383	1.68	Lee (1979);
	F	41.0	1712	1099	1.56	Leonard and Robertson (1992)
Ache (Paraguay)	M	59.6	3327	1531	2.17	Hill et al. (1984)
	F	51.8	2626	1394	1.88	Leonard and Robertson (1992)
Inuit (Canada)	M	65.0	3010	1673	1.80	Godin and
	F	55.0	2350	1305	1.80	Shephard (1973)
Pastoralists:						
Evenki (Russia)	M	58.4	2681	1558	1.75	Leonard
	F	52.7	2067	1288	1.63	(2002)
Agriculturalists:						
Aymara (Bolivia)	M	54.6	2713	1355	2.00	Kashiwazaki
	F	50.5	2376	1166	2.03	(1999)
Quichua (Ecuador)	M	61.3	3810	1601	2.38	Leonard et
	F	55.7	2460	1252	1.96	al. (1995)
Coastal Ecuador	M	55.6	2416	1529	1.58	Leonard et
	F	47.8	1993	1226	1.63	al. (1995)
Gambia	M	61.2	3848	1604	2.40	Heini et al. (1996)
	F	50.3	2500	1236	2.03	Heini et al. (1991)
Huli (PNG)	M	63.6	3138	1704	1.84	Yamauchi et
	F	53.3	2639	1391	1.88	al. (2001)
Thailand	M	55.1	2892	1322	2.20	Murayama and
	F	57.7	2218	1217	1.83	Ohtsuka (1999)
Subsistence Populations (Average):	M	58.0	3015	1525	1.98	
	F	51.6	2294	1257	1.82	
Industrial Societies:	M	70.1	2873	1659	1.73	Institute of
	F	58.6	2234	1300	1.72	Medicine (2002)

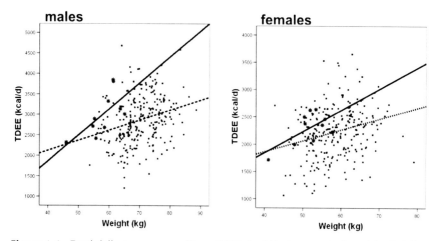

Figure 1.4. Total daily energy expenditure (TDEE; kcal/day) versus body weight (kg) for adult men and women of industrial and subsistence-level populations. Individuals from subsistence-level groups have systematically higher levels of energy expenditure at a given body weight.

Such reductions in daily energy expenditure associated with our "modern" lifestyles are a main contributor to the growing problem of obesity throughout the world. In some sense obesity and other chronic diseases of the modern world (diabetes and cardiovascular disease, for example) represent a continuation of trends that were started early in our evolutionary history. We have developed a diet that is extremely rich in calories while at the same time minimizing the amount of energy necessary for physical work and activity.

Thus, it is not simply the consumption of "bad" or "unhealthy" foods that has produced the obesity epidemic. Rather, it is changes in both diet and activity patterns that have produced ever greater surpluses of energy. The ongoing processes of change are in reality an extension of adaptive trends seen throughout human evolution—increasing the nutritional density of our diets, while reducing the time and energy associated with obtaining food. The difference now is that the changes are occurring at a much more rapid rate—producing large imbalances between "energy in" and "energy out."

Consequently, we suggest that in addressing the obesity problem and promoting better nutritional health, additional attention needs to be given to energy expenditure and activity levels. We see movement in this direction—the most recent Institute of Medicine's dietary and nutritional guidelines (2002) include recommendations on physical activity. At this point, however, there remains considerable debate about how much physical activity is necessary to promote a healthy lifestyle. This is an issue that we can and should address from a comparative, ecological, and evolutionary perspective—looking at the range of energy needs and activity levels of more traditionally living societies to give us a better sense of the level of exertion and metabolic demands for much of our evolutionary history.

Conclusion

Like other primates, we are omnivores, who can subsist on a diverse and eclectic mix of foods (e.g., Harding and Teleki 1981; Stanford 1996). However, the evolution of our disproportionately large human brains has had important and unique implications for the nutritional biology of our species. Our large brains are energetically expensive, yet, paradoxically, our overall metabolic requirements are similar to those of any comparably sized mammal. As a consequence, humans expend a relatively larger proportion of their resting energy budget on brain metabolism than other primates or nonprimate mammals.

Comparative analyses of primate dietary patterns indicate that the high costs of large human brains are supported, in part, by diets that are relatively rich in energy and other nutrients. Among living primates, the relative proportion of metabolic energy allocated to the brain is positively correlated with dietary quality. Humans fall at the positive end of this relationship, having both a very high-quality diet and a large brain.

The metabolic demands of the brain are particularly extreme during infancy and early childhood, when rapid brain growth is occurring and brain to body mass ratios are higher than in adulthood. To accommodate the very high energy demands of brain during early life, human infants are born with relatively high levels of adiposity, and continue to gain fat during the first twelve to eighteen months of life. These greater levels of body fatness and reduced levels of muscle mass allow human infants to accommodate the growth of their large brains in two important ways: (1) by having a ready supply of stored energy to "feed the brain," and (2) by reducing the total energy costs of the rest of the body.

The difficulties associated with fueling brain metabolism in infants are evident among impoverished populations of the world today. Under conditions where disease load and poor diet quality limit the availability of energy and nutrients to young children, we find that growth in length/height is severely compromised while levels of body fatness are preserved. This response allows for brain metabolism to be relatively protected by down-regulating the energy costs of body growth and preserving sufficient stores of energy as fat.

In addition to the nutritional demands of our large brains, throughout most of our evolutionary history, the acquisition of our high-quality diets required substantial expenditure of energy and movement over much larger areas than other primates. Over time, however, we have become ever-more efficient at extracting energy and nutrients from our environments. Today, daily energy demands in populations of the industrialized world are considerably less than those of subsistence-level (e.g., foraging, farming) societies. When matched for body weight, the differences in daily energy expenditure between Western and subsistence-level adults is between 300 and 900 kcal/day. This suggests that the transition from a subsistence to a sedentary, modern lifestyle is associated with a 15–30 percent reduction in one's maintenance energy needs.

In contrast to changes in energy expenditure, absolute daily energy intakes for adults of the industrialized world do not appear to differ markedly from those of subsistence-level societies. Thus, contrary to conventional wisdom, the dramatic rise of obesity in the United States and others parts of the industrialized world cannot be attributed simply to marked increases in absolute caloric consumption. Rather, it is the "imbalance" between energy intake and energy expenditure that is the root cause.

In this context, the problems of "overnutrition" currently seen worldwide are the extension of deep trends from our past. Addressing these problems will thus require attention to both the intake and expenditure sides of the energy balance equation.

References

Aiello, L. C., and P. Wheeler. 1995. The expensive-tissue hypothesis: The brain and the digestive system in human and primate evolution. *Current Anthropology* 36: 199–221.

Allen, L. H. 1994. Nutritional influences on linear growth: A general review. *European Journal of Clinical Nutrition* 48: S75–S89.

Allen, L. H., J. R. Backstrand, E. J. Stanek, G. H. Pelto, et al. 1992. Interactive effects of dietary quality on the growth and attained size of young Mexican children. *American Journal of Clinical Nutrition* 56: 353–64.

Antón, S. C. 2003. A natural history of *Homo erectus*. *Yearbook of Physical Anthropology* 46: 126–70.

Antón, S. C., and C. C. Swisher III. 2001. Evolution of cranial capacity in Asian *Homo erectus*. In *A Scientific Life: Papers in Honor of Dr. T. Jacob*, ed. E. Indriati, 25–39. Yogyakarta, Indonesia: Bigraf.

Antón, S. C., W. R. Leonard, and M. L. Robertson. 2002. An ecomorphological model of the initial hominid dispersal from Africa. *Journal of Human Evolution* 43: 773–85.

Audette, R.V., and T. Gilchrist. 2000. *Neanderthin: Eat Like a Caveman and Achieve a Lean, Strong, Healthy Body.* New York: St. Martins.

Behrensmeyer, K., N. E. Todd, R. Potts, and G. E. McBrinn. 1997. Late Pliocene faunal turnover in the Turkana basin, Kenya and Ethiopia. *Science* 278: 1589–94.

Berti, P. R., W. R. Leonard, and W. J. Berti. 1998. Stunting in an Andean community: Prevalence and etiology. *American Journal of Human Biology* 10: 229–40.

Black, A. E., W. A. Coward, T. J. Cole, and A. M. Prentice. 1996. Human energy expenditure in affluent societies: An analysis of 574 doubly-labeled water measurements. *European Journal of Clinical Nutrition* 50: 72–92.

Bobe, R., and A. K. Behrensmeyer. 2002. Faunal change, environmental variability and late Pliocene hominin evolution. *Journal of Human Evolution* 42: 475–97.

Briefel, R. R., and C. L. Johnson. 2004. Secular trends in dietary intake in the United States. *Annual Review of Nutrition* 24: 401–31.

Cordain, L., J. Brand-Miller, S. B. Eaton, N. Mann, et al. 2000. Plant to animal subsistence ratios and macronutrient energy estimations in worldwide hunter-gatherer diets. *American Journal of Clinical Nutrition* 71: 682–92.

Cordain, L., S. B. Eaton, A. Sebastian, N. Mann, et al. 2005. Origins and evolution of the Western diet: Health implications for the 21st century. *American Journal of Clinical Nutrition* 81: 341–54.

Crawford, M. A. 1992. The role of dietary fatty acids in biology: Their place in the evolution of the human brain. *Nutrition Reviews* 50: 3–11.

Crawford, M. A., and D. Marsh. 1995. *Nutrition and Evolution.* London: Keats.

Cunnane, S. C. 2005. *Survival of the Fattest: The Key to Human Brain Evolution.* Singapore: World Scientific Publishing Co.

DeMenocal, P. B. 2004. African climate change and faunal evolution during the Pliocene-Pleistocene. *Earth and Planetary Science Letters* 220: 3–24.

Dewey, K. G., M. J. Heinig, L. A. Nommsen, J. M. Peerson et al. 1993. Breast-fed infants are leaner than formula-fed infants at 1 y of age: The Darling Study. *American Journal of Clinical Nutrition* 52: 140–45.

Eaton, S. B. 2006. The ancestral human diet: What was it and should it be a paradigm for contemporary nutrition? *Proceedings of the Nutrition Society* 65: 1–6.

Eaton, S. B., and M. J. Konner. 1985. Paleolithic nutrition: A consideration of its nature and current implications. *New England Journal of Medicine* 312: 283–89.

Eaton, S. B., M. Shostack, and M. J. Konner. 1988. *The Paleolithic Prescription.* New York: Harper-Collins.

Flegal, K. M., M. D. Carroll, C. L. Ogden, and C. L. Johnson. 2002. Prevalence and trends in obesity among US adults, 1999–2000. *Journal of the American Medical Association* 288: 1723–27.

Food and Agriculture Organization, World Health Organization, and United Nations University (FAO/WHO/UNU) 1985. *Energy and Protein Requirements. Report of Joint FAO/WHO/UNU Expert Consultation.* WHO Technical Report Series No. 724. Geneva: World Health Organization.

———. 2004. *Human Energy Requirements. Report of a Joint FAO/WHO/UNU Expert Consultation.* Geneva: World Health Organization.

Foster, Z., E. Byron, V. Reyes-García, T. Huanca et al. 2005. Physical growth and nutritional status of Tsimane' Amerindian children of lowland Bolivia. *American Journal of Physical Anthropology* 126: 343–51.

Gabunia, L., A. Vekua, D. Lordkipanidze, C. C. Swisher III et al. 2000. Earliest Pleistocene cranial remains from Dmanisi, Republic of Georgia: Taxonomy, geological setting, and age. *Science* 288: 1019–25.

Gabunia, L., S. C. Antón, D. Lordkipanidze, A. Vekua et al. 2001. Dmanisi and dispersal. *Evolutionary Anthropology* 10: 158–70.

Garn, S. M., and W. R. Leonard. 1989. What did our ancestors eat? *Nutrition Reviews* 47: 337–45.

Godin, G., and R. J. Shephard. 1973. Activity patterns of the Canadian Eskimo. In *Polar Human Biology*, ed. O. G. Edholm and E. K. E. Gunderson, 193–215. Chichester, UK: Heinemann Books.

Habitch J. P., R. Martorell, and J. A. Rivera. 1995. Nutritional impact of supplementation in the INCAP longitudinal study: Analytic strategies and inferences. *Journal of Nutrition* 125: 1042S–1050S.

Harding, R. S. O., and G. Teleki. 1981. *Omnivorous Primates*. New York: Columbia University Press

Harestad, A. S., and F. L. Bunnell. 1979. Home range and body weight: A re-evaluation. *Ecology* 60: 389–402.

Harris, J. W. K., and S. Capaldo. 1993. The earliest stone tools: Their implications for an understanding of the activities and behavior of late Pliocene hominids. In *The Use of Tools by Human and Nonhuman Primates*, ed. A. Berthelet and J. Chavaillon, 196–220. Oxford: Oxford Science Publications.

Heini, A., Y. Schutz, E. Diaz, A. M. Prentice et al. 1991. Free living energy expenditure measured by two independent techniques in pregnant and non-pregnant Gambian women. *American Journal of Physiology* 261: E9–E17.

Heini, A. F., G. Minghelli, E. Diaz, A. M. Prentice et al. 1996. Free-living energy expenditure assessed by two different methods in rural Gambian men. *European Journal of Clinical Nutrition* 50: 284–89.

Hill, K. R., K. Hawkes, M. Hurtado, and H. Kaplan. 1984. Seasonal variance in the diet of Ache hunter-gatherers in Eastern Paraguay. *Human Ecology* 12: 101–35.

Holliday, M. A. 1986. Body composition and energy needs during growth. In *Human Growth: A Comprehensive Treatise*, volume 2. 2nd ed., ed. F. Falkner and J. M. Tanner, 101–17. New York: Plenum Press.

Institute of Medicine of the National Academies 2002. *Dietary Reference Intakes: Energy, Carbohydrate, Fiber, Fat, Fatty Acids, Cholesterol, Protein, and Amino Acids.* Washington, DC: National Academies Press, http://www.iom.edu/?id=15075.

Kashiwazaki, H. 1999. Heart rate monitoring as a field method for estimating energy expenditure as evaluated by the doubly labeled water method. *Journal of Nutritional Science and Vitaminology* 45: 79–94.

Kety, S. S. 1957. The general metabolism of the brain *in vivo*. In *Metabolism of the Central Nervous System*, ed. D. Richter, 221–37. New York: Pergamon.

Kleiber, M. 1961. *The Fire of Life*. New York: Wiley.

Kuzawa, C. W. 1998. Adipose tissue in human infancy and childhood: An evolutionary perspective. *Yearbook of Physical Anthropology* 41: 177–209.

Lee, R. B. 1979. *The !Kung San: Men, Women, and Work in a Foraging Society*. Cambridge: Cambridge University Press.

Leonard, W. R. 2000. Human nutritional evolution. In *Human Biology: An Evolutionary and Biocultural Approach,* ed. S. Stinson, B. Bogin, R. Huss-Ashmore, and D. O'Rourke, 295–344. New York: Wiley-Liss.

———. 2002. Food for thought: Dietary change was a driving force in human evolution. *Scientific American* 287 (6): 106–15.

Leonard, W. R., and M. L. Robertson. 1992. Nutritional requirements and human evolution: A bioenergetics model. *American Journal of Human Biology* 4: 179–95.

———. 1994. Evolutionary perspectives on human nutrition: The influence of brain and body size on diet and metabolism. *American Journal of Human Biology* 6: 77–88.

———. 1997a. Comparative primate energetics and hominid evolution. *American Journal of Physical Anthropology* 102: 265–81.

———. 1997b. Rethinking the energetics of bipedality. *Current Anthropology* 38: 304–9.

Leonard, W. R., K. M. DeWalt, J. S. Stansbury, and M. K. McCaston. 2000. The influence of dietary quality on the growth of highland and coastal Ecuadorian children. *American Journal of Human Biology* 12: 825–37.

Leonard, W. R., P. T. Katzmarzyk, M. A. Stephen, and A. G. P. Ross. 1995. Comparison of the heart rate-monitoring and factorial methods: Assessment of energy expenditure in highland and coastal Ecuador. *American Journal of Clinical Nutrition* 61: 1146–52.

Leonard, W. R., M. L. Robertson, J. J. Snodgrass, and C. W. Kuzawa. 2003. Metabolic correlates of hominid brain evolution. *Comparative Biochemistry and Physiology, Part A* 135: 5–15.

Martin, R. D. 1989. *Primate Origins and Evolution: A Phylogenetic Reconstruction.* Princeton, NJ: Princeton University Press.

Martorell, R., and J-P. Habicht. 1986. Growth in early childhood in developing countries. In *Human Growth: A Comprehensive Treatise,* volume 3. 2nd ed., ed. F. Falkner and J. M. Tanner, 241–62. New York: Plenum.

McHenry, H. M. 1992. Body size and proportions in early hominids. *American Journal of Physical Anthropology* 87: 407–31.

———. 1994a. Tempo and mode in human evolution. *Proceedings of the National Academy of Sciences (USA)* 91: 6780–86.

———. 1994b. Behavioral ecological implications of early hominid body size. *Journal of Human Evolution* 27: 77–87.

McHenry, H. M., and K. Coffing. 2000. *Australopithecus* to *Homo:* Transformations in body and mind. *Annual Reviews of Anthropology* 29: 125–46.

Murayama, N., and R. Ohtsuka. 1999. Heart rate indicators for assessing physical activity level in the field. *American Journal of Human Biology* 11: 647–57.

O'Dea, K. 1991. Traditional diet and food preferences of Australian aboriginal hunter-gatherers. *Philosophical Transactions of the Royal Society of London Biological Sciences* 334: 233–40.

Ogden, C. L., M. D. Carroll, L. R. Curtin, M. A. McDowell et al. 2006. Prevalence of overweight and obesity in the United States, 1999–2004. *Journal of the American Medical Association* 295: 1549–55.

Plummer, T. 2004. Flaked stones and old bones: Biological and cultural evolution at the dawn of technology. *Yearbook of Physical Anthropology* 47: 118–64.

Reed, K. 1997. Early hominid evolution and ecological change through the African Plio-Pleistocene. *Journal of Human Evolution* 32: 289–322.

Richard, A. F. 1985. *Primates in Nature.* New York: W. H. Freeman.

Ruff, C. B., E. Trinkaus, and T. W. Holliday. 1997. Body mass and encephalization in Pleistocene *Homo. Nature* 387: 173–76.

Sailer, L. D., S. J. C. Gaulin, J. S. Boster, and J. A. Kurland. 1985. Measuring the relationship between dietary quality and body size in primates. *Primates* 26: 14–27.

Stanford, C. B. 1996. The hunting ecology of wild chimpanzees: Implications for the evolutionary ecology of Pliocene hominids. *American Anthropologist* 98: 96–113.

Teleki, G. 1981. The omnivorous diet and eclectic feeding habits of the chimpanzees of Gombe National Park. In *Omnivorous Primates,* ed. R. S. O Harding and G. Teleki, 303–43. New York: Columbia University Press.

Ungar, P. S. 2007. *Evolution of the Human Diet: The Known, the Unknown, and the Unknowable.* New York: Oxford University Press.

Vrba, E. S. 1995. The fossil record of African antelopes relative to human evolution. In *Paleoclimate and Evolution, with Emphasis on Human Origins,* ed. E. S. Vrba, G. H. Denton, T. C. Partridge, and L. H. Burkle, 385–424. New Haven, CT: Yale University Press.

Wolpoff, M. H. 1999. *Paleoanthropology.* 2nd ed. Boston: McGraw-Hill.

Wynn, J. G. 2004. Influence of Plio-Pleistocene aridification on human evolution: Evidence from paleosols from the Turkana Basin, Kenya. *American Journal of Physical Anthropology* 123: 106–18.

Yamauchi, T., M. Umezaki, and R. Ohtsuka. 2001. Physical activity and subsistence pattern of the Huli, a Papua New Guinea highland population. *American Journal of Physical Anthropology* 114: 258–68.

• 2 •
Child Growth among Southern African Foragers in the Past

S. Pfeiffer and L. Harrington

Introduction

The survival of a stable human community requires not only successful reproduction of the next generation, but also the survival of those offspring to reproductive age. For many millennia, human ancestral communities were comprised of direct-return hunter-gatherers, sometimes also called hunter-collector-fishers, henceforth referred to here as foragers. Those foragers not only survived, but gradually became more numerous. This suggests that childhood was survivable, but can we say anything more specific about its nature?

Characteristics of foraging band societies include reliance on mobility, low population density, exploitation of seasonal food resources, and little access to food storage. Taken together, these factors can contribute to an environment of low food security, combined with few buffers that would protect children from environmental hazards. The latter can include extremes of temperature and aridity, noxious plants and animals, and predators. Seasonal food shortages can challenge child survival, from gestation onward. Persistent shortfalls in the amount or type of food energy available can lead to disruption of the normal tempo of body growth, or even permanent stunting.

On the other hand, when foragers are intimately familiar with the resources within their habitat, particularly when that range is within a relatively productive ecosystem, a child's life could be both secure and healthful. The combination of small group size and group mobility would contribute to a low infectious disease load. Diverse food sources can contribute to nutritional balance and an absence of the nutritional deficiency diseases that are often found in economies that rely on a single basic carbohydrate or "superfood." When a habitat includes food sources that can be harvested in small packages, these foods may be directly obtainable by children from a young age, allowing them to satisfy part of their own nutritional requirements, albeit not without risk.

Foraging in the Cape

The foraging environment of the South African Cape has been successfully exploited since the emergence of *Homo sapiens*. Coastal Middle Stone Age sites like Klasies River Mouth, dated to ca. 90,000 to 120,000 BP (Singer and Wymer 1982), show that people exploited shellfish and other marine resources, as well as terrestrial foods. Indeed, Klasies River Mouth has been called the oldest known "sea food restaurant" (Deacon and Deacon 1999). Other Middle Stone Age sites are located far from the coast; the tool kit and faunal remains all suggest exploitation of a wide range of foods through active hunting and family foraging groups. During the Last Glacial Maximum, the southern African interior became relatively cold and dry, and populations were attracted toward the southern coast. Cycles of aridity have affected human habitation of the interior, but rock shelters in the Cape Fold Mountains and river mouths appear to have remained attractive locales (Deacon and Deacon 1999).

During the Holocene (the last 10,000 years), the distribution of archaeological sites indicates that foraging groups most commonly exploited the coastal regions and the variable but generally plentiful resources of the *fynbos* biotic province, with fewer people exploiting the interior grasslands. The *fynbos*, literally "fine bush," is comprised of herbaceous plants like ericas, proteas and restionacae, with a low representation of grasses (Meadows and Sugden 1993). Within the *fynbos* and throughout southern Africa, various native plants produce highly nutritious underground storage organs in the form of roots, bulbs, corms, and rhizomes. Southern Africa is known for its floral diversity, said to include some 30,000 species of flowering plants, accounting for almost 10 percent of the world's higher plant types (van Wyk and Gericke 2000). These plants provide habitats for a wide range of animal life, from small rodents and tortoises through to large antelopes and elephants. There is a richly productive marine ecosystem, thanks to the mixing of the icy Benguela current of the South Atlantic and the warm Agulhas current of the southwest Indian Ocean. Marine protein, including fish, marine mammals, and mollusks were generally abundant, especially along rocky shorelines. In modern times archaeological sites are often first identified through the presence of very large shell middens.

There are multiple lines of evidence supporting an ancestor-descendant relationship between Later Stone Age foragers and historically known Khoe-San peoples of the Kalahari region.[1] If habitual behaviors affect the skeleton, hypotheses can test whether ethnographically documented Khoe-San behaviors extended back in time and into the richer environment of the Cape. Archaeological sites from ca. 40,000 BP to the historic era are categorized as Later Stone Age. The initial Holocene stone tool assemblage commonly known as Oakhurst was replaced by various microlithic assemblages in the second half of the Holocene (Deacon and Deacon 1999). All Later Stone Age assemblages include more bone, shell, and ostrich eggshell artifacts than the earlier Middle Stone Age complexes. Variation in the proportions of small scrapers, bladelets, and adzes are some of the regional distinctions upon which archaeologists base reconstructions of foraging adaptations during this time period (Mitchell 2002; for plot, see Barham and Mitchell 2008). At about 2000 BP, some pastoralism (evidence of sheep herding) appears in the region, as well as the production of pottery. The archaeological remains and the human skeletons discussed in this chapter either have chronometric dates prior to 2000 BP or their archaeological contexts are strongly consistent with known forager burial contexts.

The Holocene Later Stone Age human skeletons come from rock shelters, excavated with variable levels of expertise, and from many chance discoveries of skeletons buried in coastal sand dunes. The pattern of adult body size is short stature, gracile frame (Sealy and Pfeiffer 2000; Pfeiffer and Sealy 2006), and lean physique. The latter is deduced from the ubiquitous presence of squatting facets and evidence of knee hyperflexion in adult skeletal remains (Dewar and Pfeiffer 2004).

It is around the end of the Pleistocene (ca. 10,000 BP) that evidence of light-draw bows (and presumably poison-tipped arrows) is first found (Parkington 1998; Wadley 1998). Measures of upper arm bone strength and asymmetry indicate a gender-based division of labor in which women's activities probably focused on work with digging sticks and grinding, while men hunted (Stock and Pfeiffer 2004). Elements of the tool kit show homogeneity through the region, though there are also local patterns. Analysis of upper arm asymmetry suggests that men in the southern (mountain forest) and western (*fynbos*) regions appear to have favored spears and light-draw bows, respectively (Pfeiffer and Stock 2002; Stock and Pfeiffer 2004). This same research indicated symmetry in the upper arm strength of women throughout the Cape. This may reflect their regular use of digging sticks. As George Silberbauer writes about the *G/wi* (a modern Kalahari Khoe-San group),

> Women and girls forage each day for food plants within an 8 km radius of the campsite. A root, tuber or bulb is dug out with a pointed digging stick, which is used to break up the sand before scraping it out by hand. This method is slow but efficient, for the equipment is light, easily portable, and made of materials that are readily available and simply converted to this purpose. Furthermore, there is no call for sudden exertion: rather, there is a slow, steady expenditure of energy, which suits the small, lightly muscled women. … The work is not strenuous, but its lack of rhythm and the fre-

quent changes of position, the slow, meandering searching, and the carrying of the growing weight of the day's find make it very tiring. (Silberbauer 1981:199–200)

A long and lively debate since the 1970s has scrutinized whether Holocene foragers of the Cape exploited regions seasonally or whether they stayed within one locale for a prolonged period (Parkington 1972; Sealy and van der Merwe 1988; Sealy 1995; Sealy 1997; Parkington 2001). Stable isotope research on human skeletal tissue suggests that groups frequently delimited their ranges to either terrestrial or coastal zones. Indeed, recent research comparing the stable isotope values of neighboring communities along the southern coast indicates the maintenance of territories between ca. 4500 and 2000 BP, such that one group had access to high-quality protein provided by a seal rookery while their neighbors did not (Sealy 2006). On the other hand, research from the Eastern Cape coast suggests that inland people would visit the shoreline and leave shell middens behind (Binneman 2004/2005).

While the full range of regional food resources were probably not exploited by any one group, proteins and carbohydrates were available in variable proportions to all communities. There was seasonality in the availability, palatability, and nutritional value of the plant resources, which would have required strategic planning. Controlled burning may have been used to influence seed release and new growth (Deacon 1976). The labor involved in collecting shellfish meant that the economic return ratio was only profitable during neap (weak) tides (Binneman 2004/2005). While shellfish collection does not require particular skill or strength, and is often the domain of women and children (Meehan 1982), a diet with a large proportion of shellfish can be maintained for only limited periods (Noli and Avery 1988). There has been debate about the food safety of a diet heavily based on shellfish (cf. Noli and Avery 1988), but archaeological evidence suggests that heavy reliance on low trophic level marine protein was common among some Later Stone Age populations (Jerardino, Branch, and Navarro 2008; Mitchell 2002).

Capture of terrestrial animals was achieved through trapping, as well as hunting. Tortoise bones and carapace sections are common in middens, and could be procured by children. Fats were probably rare and highly valued. This may help to explain both the high status of large antelopes like the kudu and eland in rock art and ethnography, and the apparent competition along the South Coast for access to juvenile seals. In a stable isotope ($\delta^{13}C$ and $\delta^{15}N$) study of 127 adults, using femur length as a proxy for stature, the source of dietary protein (terrestrial versus marine) did not explain a significant amount of stature variance, although a positive correlation that was found with femoral head diameter suggests slightly greater body mass may be linked to more reliance on high trophic level marine protein (Pfeiffer and Sealy 2006). The amount of food, rather than the type of protein available, was more likely to significantly affect growth.

One ethnographically documented practice of modern-era Khoe-San foragers that is particularly relevant to the study of child growth is the practice of breastfeeding with minimal introduction of supplementary food sources until a child is well past one year of age. It could be reasoned that this pattern may be unique to the Kalahari, if the sparse food base of the Kalahari would put a younger weanling at risk. To explore the possibility of prolonged breastfeeding in a richer environment among Khoe-San ancestors, the stable nitrogen (δ^{15}N) and carbon (δ^{13}C) isotope ratios were measured from bone collagen of thirty-five infants and juveniles from a large southern coast site, Matjes River Rock Shelter (Clayton, Sealy, and Pfeiffer 2006). The isotopic values from thirty adults from the site are clustered around a lower δ^{15}N value, so the trophic effect of breast milk is apparent. Children at Matjes River Rock Shelter show consistently heightened trophic levels for at least the first 1.5 years after birth, with diminishing and increasingly variable levels thereafter. This is consistent with patterns reported for historic-era Kalahari foraging communities of prolonged on-demand breastfeeding, cessation of breast feeding by about four years (Konner 2005), and birth spacing of approximately three years (Schapera 1930; Howell 1979; Lee 1979; Shostak 1981). Hence, breast milk provided the dominant source of protein to forager infants, even when those foragers lived in an environment with abundant food resources.

Ethnographic Studies of Khoe-San Children

As part of a multidisciplinary study of the people who were then usually called the !Kung Bushmen of Botswana (now called the *Ju/'hoansi*), Nancy Howell collected cross-sectional information on the weights and statures of 165 children, virtually a complete survey of the permanent juvenile residents of the Dobe area from 1967 to 1969 (Howell 1979, 2000). She recorded a mean birth weight of 3.08 kg (s.d. 0.458), in ten newborns, with just one infant falling below the World Health Organization low birth weight criterion of 2500 g. She documented heights and weights cross-sectionally of children to twenty years of age. At the same time, Patricia Draper collected behavioral information on children's movements, play patterns, and relationships (Draper 1976). The fact of small body size among the *Ju/'hoansi* was not lost on the researchers of the Harvard Kalahari Study. Biomedical researchers documented various maladies in children of the Dobe region, but found no evidence for nutritional or infectious sources for the lessened magnitude of growth, relative to other populations (Truswell and Hanson 1976).

Draper and Howell have recently combined the information on growth and behavior for a sample of fifty-one children who were included in both of their studies (Draper and Howell 2005). Comparing the sex-specific weight, stature, and body mass index information to that of the US National Center for Health Statistics, the *Ju/'hoan* children are short and light. Their absolute weights are well below the median expectations, and their body mass indices are also below median values, although body mass values deviate to a lesser degree. The researchers explored possible correlations between the variance within the growth data and variable behaviors, including time spent in the presence of the mother or father, or other children; time spent in physical contact with someone else; time spent in low energy activity; and time spent in or near the home village. They did not find empirical support for an influence on growth coming from kinship, family composition, or children's behaviors. They conclude that area kin groups did not have differential access to resources, such that some children would be better off than others. They conclude, "The multiple cross linkages created by bilaterality, bilocal residence practices, name relationships, trading relationships, and the pervasive rule on sharing were apparently holding at the time of the study" (Draper and Howell 2005: 280). Given the existence of various behavioral linkages between the Later Stone Age people and the ethnographically documented historic communities of the Kalahari, these conclusions can be incorporated into hypotheses about the growth of Later Stone Age children.

Previous Studies of Later Stone Age Remains of Children[2]

The juvenile skeletal remains of the southern African Later Stone Age have been studied from the perspective of age at death based on dental crown formation, skeletal morphology, chronic stress, and disease indicators. A study of the pattern of relative dental crown formation confirmed that dental development follows the same pattern as that of other modern populations (Matzke 2000). This is an important assumption for the subsequent work that evaluates aspects of growth relative to dental age at death. A study of the proportions of the basiocciput (the basilar ossification center of the occipital) relative to dental age demonstrated that the basicranial shape and size in Later Stone Age children aged about six years and younger are consistent with published values for other populations (Harrington 2003).

Study of chronic and intermittent biological stress in Later Stone Age juveniles has focused on the presence of cribra orbitalia and growth arrest lines (Pfeiffer 2007). The survey included fifty-eight infants (defined in this study as children up to one year of age), children (two to twelve years) and adolescents (thirteen to eighteen years). Age at death of juveniles is based on dental formation standards (Moorrees, Fanning, and Hunt 1963a, 1963b; Smith 1991), supplemented by diaphysis lengths when teeth are not extant. In the latter case, diaphyses of Later Stone Age children with preserved teeth must be used as the reference sample, since the mean adult body size is small (Sealy and Pfeiffer 2000). Ten of thirty-six crania or partial crania show cribra orbitalia, and one of thirty-eight vaults shows signs of porotic hyperostosis. Cribra orbitalia is rare in the early years, but six of the ten children aged 6–15 years show this indication of anemia.

The femur, tibia, and radius of forty-three children have been radiographed to assess growth arrest lines at a maximum of eight locations per skeleton: distal femora, proximal and distal tibiae, and distal radii. Not surprisingly, these lines that mark resumed growth after an interruption are least common in infants of less than one year of age, where there is often no line, or no more than one line at each growing end. In the remains of seven children aged 3–6 years, most tibiae and radii show at least one line, but fewer than half of these affected bones show multiple lines. The most affected age group, with almost one-third of the children showing multiple lines at multiple sites and considerable involvement at the slow-growing distal femur site, is the group of ten children aged 6–11 years. Among the adolescents the frequency of lines is lower; only one of six shows multiple lines at any growth site. This may reflect the effect of rapid adolescent growth, erasing growth arrest lines through remodeling (Pfeiffer 2007).

The patterns seen in both the cribra orbitalia and the growth arrest line data suggest that mid-childhood may have been a vulnerable period for children. Once weaning was complete, children may have been exploring potentially risky environments, and may have been expected to provide some portion of their own food. There appears to be some association between the development of cribra orbitalia and growth arrest lines. Sixteen forager skeletons of children have preserved orbital regions and at least one of the long bones of interest preserved. Of the five of these sixteen skeletons with cribra orbitalia, three show two or more growth arrest lines at one or more bone growth sites. Of the eleven skeletons without cribra orbitalia, just two show two or more growth arrest lines at one or more growth sites.

To date, only one instance of chronic ill health has been found in a Later Stone Age child. The skeletal remains of an infant (ca. 4–5 months of age) from a southwest South African rock shelter at Byneskranskop (SAM-AP 6060) show pervasive abnormalities that are consistent with the effects of rickets associated with an inborn error of metabolism. Diagnostic features include beading of the costochondral junctions of the ribs, flaring and tilting of the metaphyses, and cupping of the distal ulna, as well as general skeletal hypertrophy. Porous thickening is especially noteworthy in the frontal-orbital region of the skull. It is described in detail elsewhere (Pfeiffer and Crowder 2004). With an uncalibrated AMS radiocarbon date of 4820 +/- 90 BP (TO-9531), this is a very early instance of this condition, among foragers whose environment and diet preclude shortages of active vitamin D or dietary calcium. The archaeological context and red ochre staining on the frontal bone shows that this infant, who likely had shown health problems since birth, was buried in a manner like that of other deceased group members.

The consistency of short stature throughout the Holocene, combined with allometric pelvic modeling that results in maintenance of the size of the obstetric canal (Kurki 2005), suggest a genetic predisposition to smallness in the foragers of southern Africa. It is improbable that the observed pattern of body size is due to pervasive nutritional deficiencies over several millennia and hundreds of kilometers of coastline. Nevertheless, the presence of nonspecific skeletal lesions in the remains of the children and the small body size among the adults necessitates attention to the process of growth in this population. Both the magnitude of linear growth and the tempo of growth are of interest. Prior case studies of some juvenile skeletons indicate that linear dimensions of the postcrania are smaller relative to dental age than is seen in other populations (Jerardino 1998; Sealy et al. 2000; Pfeiffer and van der Merwe 2004), but the cross-sectional distance curves appear normal in their shape.

The attainment of adult stature through linear growth is the end product of a complex trajectory of growth. Linear growth normally follows an s-shaped curve, with more rapid growth during infant and toddler years, slower mid-childhood growth, then an adolescent growth spurt. The manner in which each child grows reflects the "canalization" of a genetic predisposition interacting with environmental factors (Waddington 1957; Tanner 1978). Small adult body size could be a product of episodic or prolonged growth stunting, that is, inhibition of magnitude and abnormal tempo. Alternately, it could result from growth following a fundamentally normal pattern, but with the magnitude of linear growth following a different genetic path. The pattern of growth that can be deduced from the available sample of southern African juvenile foragers can help us assess the alternatives.

Analysis of Growth

This analysis proceeds in two parts. Cranial growth relative to dental maturation will be explored through information from the cranial base. Postcranial growth will be summarized in tabular form but the length of the growing femur will be the focus of the analysis, so that the tempo and magnitude of Later Stone Age linear growth can be compared to that of other populations. The skeletons included in the sample are those of forager children from coastal (rock shelter and sand dune) sites distributed throughout the South African Cape (Fig. 2.1, adapted from Churchill and Morris 1998). Juveniles are defined as individuals with at least one open long bone epiphysis. Most skeletons have radiocarbon dates earlier than 2000 BP, or they are closely associated with other skeletons that are dated to this timeframe. A small number have radiocarbon dates that are more recent than 2000 BP, but are included because their archaeological contexts are consistent with forager burial customs. Measurements were made collaboratively by a small number of researchers, all of whom followed standard osteometric protocols (Fazekas and Kosa 1978; Buikstra and Ubelaker 1994). The analysis of postcranial linear growth is focused on femoral length, taken as a proxy for stature. Where at least one femoral epiphysis is unfused, the measurements reported estimate diaphyseal length; where all epiphyses had begun to fuse, the measurements reported estimate maximum femoral length.

Any analysis of growth based on an archaeologically derived skeletal sample is affected, in no small way, by the techniques employed for estimation of chronological age and the subsequent classification of individuals into age cohorts (Saunders et al. 1993; Saunders, Hoppa, and Southern 1993). These issues are at play, particularly, in comparative analyses of patterns of growth between populations; therefore, the meth-

Child Growth among South African Foragers in the Past • 43

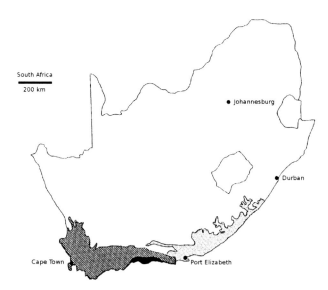

Figure 2.1. The areas from which Later Stone Age skeletons have been recovered, within fynbos (striped), forest (solid) and savanna (stippled) ecosystems along the southern coast of South Africa. Adapted from Churchill and Morris 1998.

ods of previous researchers must be accounted for before looking to any biological explanation for apparent differences observed in skeletal growth profiles (Saunders, Hoppa, and Southern 1993; Humphrey 2003). The following presentation of data by age cohorts is for the purpose of comparison with other populations. The selection of age groupings for cohorts is arbitrary, and follows the methods of previous researchers.

Age estimates for Later Stone Age juveniles are based on dental formation and eruption wherever dental remains were preserved (Moorrees, Fanning, and Hunt 1963a, 1963b; Ubelaker 1989; Smith 1991). Individual age estimates for all preserved teeth were included to form an average estimated dental age for each individual. In cases where no teeth were preserved, age at death was assessed based on degree of epiphyseal fusion using dentally aged individuals from within the sample as the standard.

Basicranial Size and Tempo of Growth

The dimensions of the basiocciput have been employed in estimating age at death in infants and young children (Redfield 1970). The relative proportion of the width and length of this ossification center have been shown to change predictably such that the ratio of width to length can be used to distinguish between younger and older fetuses, and younger and older infants (Redfield 1970; Scheuer and MacLaughlin-Black 1994). The basiocciput fuses to the lateral ossification centers of the occipital at about six years of age. The data for basiocciput age estimation can also be used to examine patterns of growth in the basicranium. Published basiocciput dimensions are available from three comparative populations, two European (Redfield 1970; Scheuer and MacLaughlin-Black 1994) and one Egyptian (Tocheri and Molto 2002). Later Stone Age adults, like their Khoe-San descendants, have estimated statures that fall at the smallest end of the range for modern human adults (Sealy and Pfeiffer 2000), yet the dimensions of the pelvic canal (Kurki 2005) suggest that Later Stone Age infants may not have been particularly small, at least insofar as the maximum dimension of the canal reflects cranial size in neonates. This would be consistent with the Kalahari birth weights recorded by Howell (1979, 2000), presented earlier.

Values for the mean maximum width of the basiocciput by age cohort are given in Table 2.1. Mistihalj is a sample of thirty-seven children excavated from the Trebisnjica river valley and dated to 1400–1475 CE (Redfield 1970) with age estimates based on dental development (Anderson 1962) and epiphyseal fusion (Krogman 1962). The Spitalfields sample consists of forty-six children with documented ages

Table 2.1. Maximum width (in millimeters) of the basiocciput in Later Stone Age juveniles and three other groups. Age is the mean age for the cohort in the LSA (Harrington 2003) and Spitalfields (Scheuer and McLaughlin-Black 1994) samples; age is the midpoint for the cohort in Dakleh (Tocheri and Molto 2002) and Mistihalj (Redfield 1970) samples.

Age (years)	LSA		Dakleh		Mistihalj		Spitalfields	
	n	Width (mm)	n	Width (mm)	n	Width (mm)	n	Width (mm)
Birth	4	15	2	26	4	17	5	15
0.25	5	17	1	15	6	18	2	15
0.50	4	18	8	19	4	22	1	18
0.75			3	20			5	21
1.00	4	17			4	23	4	21
1.50	3	22	9	22	5	24	8	22
2.00	1	27					6	24
2.50					6	27		
3.00			2	28			9	25
3.50	1	32						
4.00	3	23	1	28			5	27
4.50					5	28	1	26
5.00	3	22	1	25				
5.50	2	23						
6.00	1	21	2	25	3	30		

at death, who were interred in a British church crypt between 1729 and 1857 CE (Molleson and Cox 1993; Scheuer and MacLaughlin-Black 1994). Dakleh is an Egyptian comparative group comprised of twenty-nine children from a cemetery site (Tocheri and Molto 2002). The basiocciput was preserved in thirty-one Later Stone Age children, and age estimation followed the method described above.

Figure 2.2 is a growth profile for basiocciput width plotted by cohorts. Values for Later Stone Age children fall within the range of available comparative populations. If the width of the basiocciput can be taken as a proxy for basicranial size, the crania of Later Stone Age infants and young children appear to be no smaller than the comparative groups to any significant degree. Moreover, each of the four groups appears to maintain approximately the same trajectory of growth in the basicranial region.

Postcranial Linear Dimensions and Tempo of Growth

Postcranial measurements for ninety-four Later Stone Age juveniles are reported. The age distribution for the sample is outlined in Table 2.2, with mean and median cohort age values provided. The following analysis focuses on the maximum length of

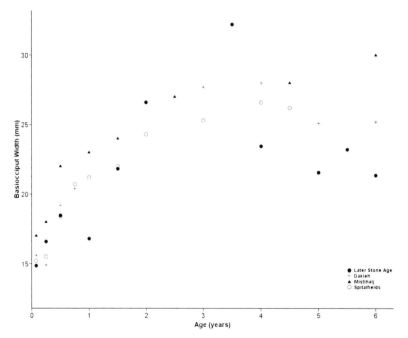

Figure 2.2. Basicranial size for age in Later Stone Age juveniles and three comparative groups.

Table 2.2. Estimated age at death and femur lengths for the Later Stone Age sample.

Cohort*	Sample Age Distribution			Femur Length			
	n	Mean Age	Median Age	n	Mean	s.d.	95% CI
0.25	18	0.26	0.25	10	86.03	9.4	5.8
0.75	5	0.55	0.57	5	107.73	14.9	13.0
1.50	7	1.30	1.42	5	123.97	14.9	13.1
2.50	4	2.58	2.48	4	160.92	15.1	14.8
3.50	2	3.63	3.63	2	186.83	21.5	29.7
4.50	5	4.75	4.60	5	188.00	11.4	10.0
5.50	10	5.50	5.63	8	213.91	12.5	8.6
6.50	8	6.43	6.43	7	238.06	12.6	9.4
7.50	4	7.39	7.37	2	233.50	26.2	36.3
9.50	2	9.39	9.39	2	297.00	17.7	24.5
10.50	1	10.45	10.45	1	276.25		
11.50	2	11.30	11.30	2	345.75	0.4	0.5
12.50	5	12.20	12.27	5	303.10	18.5	16.3
13.50	2	13.26	13.26	1	356.50		
15.50	6	15.63	15.90	6	372.33	23.8	19.0
16.50	3	16.65	16.53	2	396.75	25.8	35.8
17.50	10	17.00	17.05	9	394.64	28.2	18.4
Total	94			76			

*Cohort values are the midpoint of the cohort.

the femoral diaphysis in seventy-six Later Stone Age juveniles with estimated ages ranging from newborn to eighteen years.

The relationship of individual femur lengths relative to estimated ages at death shows a steady increase in femur length in the cross-sectional sample, with no indication of periods of slowed growth; indeed growth has seemingly not yet ceased by age eighteen.

The values from Later Stone Age children can be compared to those from the Denver growth study (Maresh 1970). The Denver growth study is a radiographic mixed longitudinal study of children aged between two months and eighteen years, who resided in Denver, Colorado between 1935 and 1967 (Maresh 1943, 1955). Values used from the Denver children are the median of the male and female mean femoral diaphysis lengths for each age cohort. Standard deviations (+/-1 and 2) were also calculated based on the median of the male and female reported standard deviation for each cohort. Later Stone Age children are consistently smaller in magnitude of femur length than the modern North American sample, often by values exceeding two standard deviations (for plot, see Harrington and Pfeiffer 2008).

The trajectory for growth begins at approximately the same place for these two groups (with neonatal femur lengths in the range of 75mm), but the Later Stone Age sample falls below two standard deviations relative to Denver children by the age of two years, and continues on a course of significantly smaller femur lengths for estimated age.

The trajectory for femur length in Later Stone Age children, while small in magnitude relative to modern North American children, is consistent in tempo. At no point in the growth profile do Later Stone Age children appear to fall off the trajectory, as would be apparent if the children were experiencing growth stunting and/or subsequent catch-up growth. Figure 2.3 illustrates the relative growth trajectories for Later Stone Age and modern North American children expressed as a percentage of

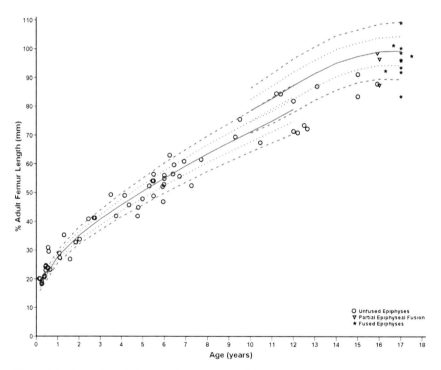

Figure 2.3. Femur length in Later Stone Age (LSA) juveniles expressed as a percentage of the mean value for femur length in LSA adults, plotted to compare with the percentage of adult length achieved in Denver growth study participants. Solid lines indicate the Denver mean, dashed lines indicate +/- 1 and 2 standard deviations.

the mean femur length for adult males and females of their respective populations. The mean femur length for Later Stone Age adults was calculated from values in Pfeiffer and Sealy (2006) (N=59), whereas the North American adult mean femur length was calculated from Maresh's (1970) values for the oldest adolescents in the Denver study (the median for eighteen-year-old males and seventeen-year-old females); this technique follows Humphrey (2000, 2003). Expressed in terms of the population-specific adult "end-point," Later Stone Age children achieve femur length outcomes that are similar to North American children, consistently falling within the range of two standard deviations of mean values for the percentage of adult femur length in the Denver sample. In the earliest ages, several Later Stone Age children are more than two standard deviations above percentage of adult length values achieved in Denver children, suggesting that North American children are growing at a somewhat slower pace relative to Later Stone Age children during this phase of growth.

The relative characteristics of growth tempo can be compared more closely by plotting the percentage of adult femur length achieved in each Later Stone Age child as a residual relative to the Denver values, and this approach can be expanded to include other available datasets (Fig. 2.4). To facilitate comparison with individual age estimates for the skeletons of past children, curves were fit to the Denver values in order to predict values for the percentage of adult femur length achieved at ages intermediate to the published cohort midpoints. This technique is described in Humphrey (2000, 2003). The residual difference in percentage of adult femur length achieved can be plotted for each child relative to the standard of the Denver growth study. Most Later Stone Age children fall within two standard deviations of the North American reference. The femur lengths of Later Stone Age children demonstrate no pattern of altered growth outcomes relative to the Denver benchmark at any particular age. Moreover, the scatter of points in this death assemblage would suggest that forager children who died were not necessarily those who were short for age; equal numbers of children exceed the Denver benchmark by two standard deviations as do those who fall under it by the same measure.

This technique for assessing relative tempo of growth in terms of a population-specific standard for the adult "end-point" is extended to five additional comparative groups for which femur length data are available for both adults and children, and whose studies employ dental age estimation methods. Table 2.3 outlines some characteristics of these comparative groups. Data points in Figure 2.4 are the mean percentage of adult femur length that was achieved by each age cohort, with interpolation lines intended to illustrate the tempo of growth for each population. Fluctuations in the plotted line indicate a deviation in the tempo of growth relative to Denver values. A population whose tempo of growth was similar to the North American reference would be illustrated by a horizontal line.

Figure 2.4 illustrates that Later Stone Age children are approaching the adult end point, on average, more quickly than North American children before the age of four years. After age three there is a marked slowing of growth tempo. One explana-

Child Growth among South African Foragers in the Past • 49

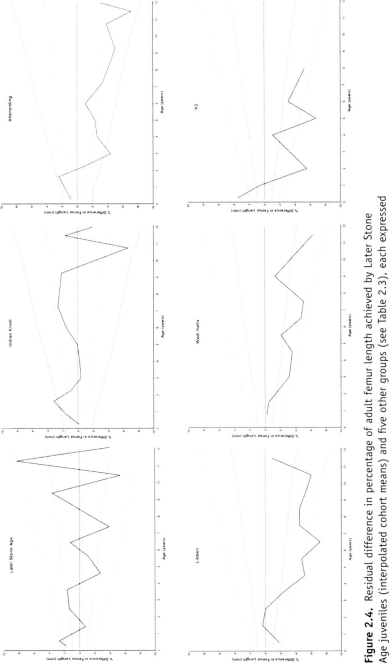

Figure 2.4. Residual difference in percentage of adult femur length achieved by Later Stone Age juveniles (interpolated cohort means) and five other groups (see Table 2.3), each expressed relative to predicted values taken from the Denver growth study. The solid line indicates the Denver mean, dashed lines indicate +/− 1 and 2 standard deviations.

Table 2.3. Comparative groups included in the analysis of linear growth.

Population	Economy	Location	Time Period	Data Sources
Alternerding	Not described	Near Munich, Germany	500–600 CE	Sundick 1978
Indian Knoll	Direct-return foraging	Kentucky, U.S.A.	ca. 5000 BP	Johnston 1962; Sundick 1978
K2	Pastoralist/ horticul-turalist	Limpopo Province, South Africa	1000–1200 CE	Steyn and Henneberg 1996
Libben	Delayed-return foraging	Ohio, U.S.A.	800–1100 CE	Lovejoy 1985; Lovejoy, Russell, and Harrison 1990
Wadi Halfa	Sedentary agriculture and urban	Sudanese Nubia	350 BCE– 1400 CE	Armelagos et al. 1972

tion for the change in growth tempo at this age may be differences in weaning times between Later Stone Age and North American children, with forager children being weaned later, by approximately age four (Konner 2005). If weaning stress is expressed by a slowed growth tempo in young children, femur lengths in the forager infants would exceed those of North American children, but show a slowed tempo following the age of weaning. After four years of age, the femur lengths of Later Stone Age children fluctuate about the mean for Denver children, again suggesting that femur length provides no insights into explaining the survivorship or mortality of forager children.

A similar pattern is observed in the Indian Knoll (Sundick 1978) sample. The tempo of growth in Indian Knoll children is also rapid in relation to the Denver reference, particularly during infancy and early childhood. Like many Later Stone Age sites, Indian Knoll bears evidence that its occupants consumed large amounts of shellfish (Johnston 1962), and prolonged breastfeeding has been postulated for this population, based on stable isotopes (Schurr and Powell 2005). Of the comparative groups included in this analysis, the economy of the Indian Knoll population may be most similar to that of Later Stone Age foragers.

The tempo of growth in the Altenerding (Sundick 1978), Wadi Halfa (Armelagos et al. 1972), and Libben (Lovejoy, Russell, and Harrison 1990) samples generally falls off quite early from the North American standard, and remains quite slowed until at least twelve years of age. It seems likely that these groups would undergo an adolescent growth spurt of significant magnitude in order to achieve the adult femur length values, if the immature femora that were measured are representative of children in those populations. The K2 (Steyn and Henneberg 1996) skeletons (n=19) represent Iron Age black farmers who lived in villages with their cattle herds. In a comparison of the juveniles from K2 to juveniles from one large Later Stone Age site (Oakhurst), Steyn (1997) found lower frequencies of cribra orbitalia and growth arrest lines at K2. The K2 children show a slowed tempo of growth that falls below two standard deviations relative to the North American standard, and is among the slowest rates of growth in the comparative groups.

Conclusion

In the context of their study of Kalahari child growth, Draper and Howell (2005: 265) noted that "It may be advantageous to be short and light when your life consists of hunting and gathering in a hot environment. But the issue deserves more careful assessment." Their perspective has merit. This analysis has focused on the tempo and magnitude of growth among Holocene foragers of the South African Cape, among communities who represent the ancestors of those Kalahari hunter-gatherers. Results suggest that the Holocene forager children were not small at birth relative to modern standards, and that they followed a pattern of linear growth that was similar to modern, healthy populations (like the Denver sample) in their attainment of adult stature.

The comparison of basiocciput dimensions indicates that cranial size is not diminished in the early years. This is consistent with an expectation that birth weights were not particularly low, and that the allometric modeling of the obstetric canal that has been previously demonstrated was adaptive.

There is debate, within the context of the "osteological paradox" (Wood et al. 1992), regarding whether signs of nonspecific stress, such as growth arrest lines and cribra orbitalia, represent distress or adaptation among affected individuals. This study documented normal growth among a relatively large number of juvenile foragers, about one-third of whom show cribra orbitalia, and almost that many show growth arrest lines. Nonspecific stress indicators are reported to be much less common within the South African Iron Age K2 sample, where the environment is similar (although not identical). Growth among the K2 children appears to be compromised. While sample size from K2 is small, this comparison suggests that the nonspecific stress indicators may, paradoxically, be symptomatic of children who were growing well in a challenging yet dietarily sufficient environment.

This study used the length of the femur to explore general linear body growth. It compared the growth of Later Stone Age forager children to children from various other subsistence strategies and ecological zones. Later Stone Age children do not appear to experience episodes of significant growth stunting in the course of becoming relatively short-statured adults. While small in absolute size, growth of the Later

Stone Age children demonstrates a pattern in tempo that is consistent with healthful growth. This pattern suggests that this Holocene population was well adapted to the particular challenges of their environment. One aspect of this adaptation may have been small body size.

Acknowledgements

The authors thank Louise Humphrey for assistance with constructing the Denver growth study comparative plots, and Maryna Steyn for making available data for K2 adults. We also thank the curators who provided access to the collections that form the basis of the research. They include: Lita Webley and Johan Binneman (Albany Museum, Grahamstown), James Brink (Florisbad Research Station, National Museum, Bloemfontein), Alan Morris (University of Cape Town), and Graham Avery and Sarah Wurz (Iziko Museums of Cape Town). The research has been financially supported by the Social Sciences and Humanities Research Council of Canada.

Notes

1. As contemporary authors attempt to be sensitive to the wishes of Khoe and San descendant groups, variations on nomenclature are proposed with some regularity. Sometimes proposals for new terms are accompanied by negative interpretations of prior terminology. This paper will follow Crawhall (2006), thus using terms that differ from the senior author's previous publications. Depending on the archaeological theory followed, the ancestors described herein may be more appropriately characterized simply as San. Whatever the nomenclature, the intent is to acknowledge the right of descendants to their voice in the matter.
2. The terms *child* and *children* are used to refer to the general category of "non-adults" in preference to the pejorative term "subadult" (Lewis 2007:2). Where specific age subgroups are discussed, the terms *infant, child,* and *adolescent* are employed.

References

Anderson, J. E. 1962. *The Human Skeleton: A Manual for Archaeologists.* Ottawa, National Museums of Canada.
Armelagos, G. J., J. H. Mielke, K. H. Owen, D. P. Van Gerven et al. 1972. Bone growth and development in prehistoric populations from Sudanese Nubia. *Journal of Human Evolution* 1 (1): 89–119.
Barham L. and P. Mitchell. 2008. *The First Africans: African Archaeology from the Earliest Toolmakers to the Most Recent Foragers.* Cambridge: Cambridge University Press.
Binneman, J. 2004/2005. Archaeological research along the southern-eastern Cape Coast part 1: Open-air shell middens. *Southern African Field Archaeology* 13/14: 49–77.

Buikstra, J. E., and D. H. Ubelaker, eds. 1994. *Standards for Data Collection from Human Skeletal Remains.* Fayetteville: Arkansas Archeological Survey.

Churchill, S. E. and A. G. Morris. 1998. Muscle marking morphology and labour intensity in prehistoric Khoisan foragers. *International Journal of Osteoarchaeology* 8: 390–411.

Clayton, F., J. Sealy, and S. Pfeiffer. 2006. Weaning age among foragers at Matjes river rock shelter, South Africa, from stable nitrogen and carbon isotope analyses. *American Journal of Physical Anthropology* 129 (2): 311–17.

Crawhall, N. 2006. Languages, genetics and archaeology: Problems and the possibilities in Africa. In *The Prehistory of Africa,* ed. H. Soodyall, 109–24. Johannesburg: Jonathan Ball Publishers.

Deacon, H. J. 1976. *Where Hunters Gathered: A Study of Holocene Stone Age People in the Eastern Cape.* Cape Town: South African Archaeological Society.

Deacon, H. J., and J. Deacon. 1999. *Human Beginnings in South Africa: Uncovering the Secrets of the Stone Age: Uncovering the Secrets of the Stone Age.* Cape Town: David Phillip Publishers.

Dewar, G., and S. Pfeiffer. 2004. Postural behavior of Later Stone Age people in South Africa. *South African Archaeological Bulletin* 59: 52–58.

Draper, P. 1976. Social and economic constraints on child life among the !Kung. In *Kalahari Hunter-Gatherers: Studies of the !Kung San and Their Neighbors,* ed. R. B. Lee and I. DeVore, 199–217. Cambridge, MA: Harvard University Press.

Draper, P. and N. Howell. 2005. The growth and kinship resources of Ju/'hoansi children. In *Hunter-Gatherer Childhoods: Evolutionary, Developmental and Cultural Perspectives,* ed. M. E. Hewlett and B. S. Lamb, 262–81. New Brunswick, NJ: Aldine Transaction.

Fazekas, I. G., and F. Kosa. 1978. *Forensic Fetal Osteology.* Budapest: Akademiai Kiado.

Harrington, L. 2003. Estimating Age at Death from Basiocciput Osteometrics: A Test of the Method in Juvenile Later Stone Age Foragers. Masters Paper, University of Toronto.

Harrington, L. and S. Pfeiffer. 2008. Juvenile mortality in southern African archaeological contexts. *South African Archaeological Bulletin* 63: 95–101.

Howell, N. 1979. *Demography of the Dobe !Kung.* New York: Academic Press.

———. 2000. *Demography of the Dobe !Kung.* New Brunswick, NJ: Aldine Transaction.

Humphrey, L. 2000. Growth studies of past populations: An overview and an example. In *Human Osteology: In Archaeology and Forensic Science,* ed. M. Cox and S. Mays, 23–38. Cambridge: Cambridge University Press.

———. 2003. Linear growth variation in the archaeological record. In *Patterns of Growth in the Genus Homo,* ed. J. L. Thompson, G. E. Krovitz and A. J. Nelson, 144–69. Cambridge: Cambridge University Press.

Jerardino, A. 1998. Excavations at Pancho's Kitchen Midden, Western Cape coast, South Africa: Further observations into the Megamidden period. *South African Archaeological Bulletin* 53: 16–25.

Jerardino, A., G. M. Branch, and R. Navarro. 2008. Human impact on precolonial west coast marine environments of South Africa. In *Human Impacts on Ancient Marine Ecosystems:*

a Global Perspective, ed. T.C. Rick and J. M. Erlandson, 279–96. Berkeley: University of California Press.

Johnston, F. E. 1962. Growth of the long bones of infants and young children at Indian Knoll. *American Journal of Physical Anthropology* 20 (3): 249–54.

Konner, M. 2005. Hunter-gather infancy and childhood: The !Kung and others. In *Hunter-Gatherer Childhoods: Evolutionary, Developmental and Cultural Perspectives,* ed. M. E. Hewlett and B. S. Lamb, 19–64. New Brunswick, NJ: Aldine Transaction.

Krogman, W. 1962. *The Human Skeleton in Forensic Medicine.* Springfield, IL: Charles C. Thomas.

Kurki, H. K. 2005. Adaptive Allometric Modeling of the Pelvis in Small-bodied Later Stone Age (Holocene) Foragers from Southern Africa. PhD thesis, University of Toronto.

Lee, R. B. 1979. *The !Kung San: Men, Women and Work in a Foraging Society.* Cambridge: Cambridge University Press.

Lewis, M. E. 2007. *The Bioarchaeology of Children: Perspectives from Biological and Forensic Anthropology.* Cambridge: Cambridge University Press.

Lovejoy, O. C., K. F. Russell, and M. L. Harrison. 1990. Long bone growth velocity in the Libben population. *American Journal of Human Biology* 2 (5): 533–41.

Maresh, M. M. 1943. Growth of major long bones in healthy children: A preliminary report on successive roentgenograms of the extremities from early infancy to twelve years of age. *American Journal of Diseases of Children* 66 (3): 227–57.

———. 1955. Linear growth of long bones of extremities from infancy through adolescence; continuing studies. *American Journal of Diseases of Children* 89 (6): 725–42.

———. 1970. Measurements from roentgenograms. In *Human Growth and Development,* ed. M. M. Maresh and R. W. McCammon, 157–200. Springfield, IL: Charles C. Thomas.

Matzke, L. 2000. Dental Development and Maturation of Juvenile Dentitions from the Later Stone Age, South Africa. Masters Paper, University of Toronto.

Meadows, M. E., and J. M. Sugden. 1993. The late quaternary palaeoecology of a floristic kingdom: the southwestern Cape South Africa. *Palaeogeography, Palaeoclimatology, Palaeoecology* 101 (3–4): 271–81.

Meehan, B. 1982. *Shell Bed to Shell Midden.* Canberra: Australian Institute of Aboriginal Studies.

Mitchell, P. 2002. *The Archaeology of Southern Africa.* Cambridge: Cambridge University Press.

Molleson, T., and M. Cox 1993. *The Spitalfields Project.* London: Council for British Archaeology.

Moorrees, C. F., E. A. Fanning, and E. E. Hunt. 1963a. Age variation of formation stages for ten permanent teeth. *Journal of Dental Research* 42: 1490–1502.

———. 1963b. Formation and resorption of three deciduous teeth in children. *American Journal of Physical Anthropology* 21: 205–13.

Noli, D., and G. Avery. 1988. Protein poisoning and coastal subsistence. *Journal of Archaeological Science* 15 (4): 395–401.

Parkington, J. 1972. Seasonal mobility in the Late Stone Age. *African Studies* 31: 151–58.

———. 1998. Resolving the past: Gender in the archaeological record of the Western Cape. In *Gender in African Prehistory,* ed. S. Kent, 25–38. Walnut Creek, CA: Altamira Press.

———. 2001. Mobility, seasonality and southern African hunter-gatherers. *South African Archaeological Bulletin* 56: 1–7.

Pfeiffer, S. 2007. The health of foragers: People of the Later Stone Age, southern Africa. In *Ancient Health: Skeletal Indicators of Agricultural and Economic Intensification,* ed. M. N. Cohen and G. M. M. Crane-Kramer, 223–36. Gainesville: University Press of Florida.

Pfeiffer, S., and C. Crowder. 2004. An ill child among mid-Holocene foragers of Southern Africa. *American Journal of Physical Anthropology* 123 (1): 23–29.

Pfeiffer, S., and N. J. van der Merwe. 2004. Cranial injuries to Later Stone Age children from the Modder River Mouth, Southwestern Cape, South Africa. *South African Archaeological Bulletin* 59: 59–64.

Pfeiffer, S., and J. Sealy. 2006. Body size among Holocene foragers of the Cape Ecozone, southern Africa. *American Journal of Physical Anthropology* 129 (1): 1–11.

Pfeiffer, S., and J. T. Stock. 2002. Upper limb morphology and the division of labor among southern African Holocene foragers. *American Journal of Physical Anthropology* S34: 124.

Redfield, A. 1970. A new aid to aging immature skeletons: Development of the occipital bone. *American Journal of Physical Anthropology* 33 (2): 207–20.

Saunders, S., C. Devito, A. Herring, R. Southern et al. 1993. Accuracy tests of tooth formation age estimations for human skeletal remains. *American Journal of Physical Anthropology* 92 (2): 173–88.

Saunders, S., R. Hoppa, and R. Southern. 1993. Diaphyseal growth in a nineteenth century skeletal sample of subadults from St Thomas' church, Belleville, Ontario. *International Journal of Osteoarchaeology* 3 (4): 265–81.

Schapera, I. 1930. *The Khoisan Peoples of South Africa.* London: Routledge and Kegan Paul.

Scheuer, L., and S. McLaughlin-Black. 1994. Age estimation from the pars basilaris of the fetal and juvenile occipital bone. *International Journal of Osteoarchaeology* 4 (4): 377–80.

Schurr, M. R. and M. L. Powell. 2005. The role of changing childhood diets in the prehistoric evolution of food production: an isotopic assessment. *American Journal of Physical Anthropology* 126: 278–94.

Sealy, J. C. 1995. Analysing the past: Modern analytical techniques, notably involving the measurement of isotope ratios, help clarify early mysteries regarding diet and trading patterns. *South African Journal of Science* 91: 11–12.

———. 1997. Stable carbon and nitrogen isotope ratios and coastal diets in the Later Stone Age of South Africa: A comparison and critical analysis of two data sets. *Ancient Biomolecules* 1: 131–47.

———. 2006. Diet, mobility, and settlement pattern in Holocene South Africa. *Current Anthropology* 47 (4): 569–95.

Sealy, J. C., and N. J. van der Merwe. 1988. Social, spatial and chronological patterning in marine food use as determined by delta13C measurements of Holocene human skeletons from the south-western Cape, South Africa. *World Archaeology* 20 (1): 87–102.

Sealy, J. C., and S. Pfeiffer. 2000. Diet, body size, and landscape use among Holocene people in the Southern Cape, South Africa. *Current Anthropology* 41 (4): 641–55.
Sealy, J. C., S. Pfeiffer, R. Yates, C. Willmore et al. 2000. Hunter-gatherer child burials from the Pakhuis Mountains, Western Cape: Growth, diet and burial practices in the Late Holocene. *South African Archaeological Bulletin* 55: 32–43.
Shostak, M. 1981. *Nisa: The Life and Words of a !Kung Woman*. Cambridge, MA: Harvard University Press.
Silberbauer, G. B. 1981. *Hunter and Habitat in the Central Kalahari Desert*. Cambridge: Cambridge University Press.
Singer, R., and J. Wymer 1982. *The Middle Stone Age at Klasies River Mouth in South Africa*. Chicago: University of Chicago Press.
Smith, B. H. 1991. Standards of human tooth formation and dental age assessment. In *Advances in Dental Anthropology*, ed. M. A. Kelley and C. S. Larsen, 143–68. New York: Wiley-Liss.
Steyn, M. 1997. A reassessment of the human skeletons from K2 and Mapungubwe (South Africa). *South African Archaeological Bulletin* 52: 14–20.
Steyn, M., and M. Henneberg. 1996. Skeletal growth of children from the Iron Age site at K2 (South Africa). *American Journal of Physical Anthropology* 100 (3): 389–96.
Stock, J. T., and S. K. Pfeiffer. 2004. Long bone robusticity and subsistence behavior among Later Stone Age foragers of the forest and fynbos biomes of South Africa. *Journal of Archaeological Science* 31 (7): 999–1013.
Sundick, R. 1978. Human skeletal growth and age determination. *HOMO: Journal of Comparative Human Biology* 29: 228–49.
Tanner, J. M. 1978. *Foetus into Man: Physical Growth from Conception to Maturity*. Cambridge, MA: Harvard University Press.
Tocheri, M. W., and E. J. Molto. 2002. Aging fetal and juvenile skeletons from Roman period Egypt using basiocciput osteometrics. *International Journal of Osteoarchaeology* 12 (5): 356–63.
Truswell, A., and J. Hanson. 1976. Medical research among the !Kung. In *Kalahari Hunter-Gatherers: Studies of the !Kung San and Their Neighbors*, ed. R. B. Lee and I. DeVore, 166–94. Cambridge, MA: Harvard University Press.
Ubelaker, D. H. 1989. *Human Skeletal Remains: Excavation, Analysis, Interpretation*. Washington, DC: Taraxacum.
Waddington, C. 1957. *The Strategy of the Genes*. London: Allen and Unwin.
Wadley, L. 1998. The invisible meat providers: Women in the Stone Age of South Africa. In *Gender in African Prehistory*, ed. S. Kent, 69–82. Walnut Creek, CA: Altamira Press.
Wood, J. W., G. R. Milner, H. C. Harpending, and K. M. Weiss. 1992. The osteological paradox: Problems of inferring prehistoric health from skeletal samples. *Current Anthropology* 33 (4): 343–70.
van Wyk, B., and N. Gericke 2000. *People's Plants: A Guide to Useful Plants of Southern Africa*. Pretoria: Briza Publications.

• 3 •

Infant and Young Child Feeding in Human Evolution

D. W. Sellen

Introduction

This chapter calls for a reconsideration of the central importance of both physiological and behavioral strategies for feeding infants and young children during human evolution and their implications for the health of mothers and babies in contemporary societies. It develops a hypothesis that the evolutionary biology of human lactation and the evolutionary ecology of complementary feeding contribute to contemporary challenges in maternal and child nutrition and health. Evolutionary anthropological and ethnographic studies are used to develop a general conceptual framework for understanding prehistoric, historic, and contemporary variation in infant and young child feeding patterns. Let us begin our discussion of the evolutionary significance of human lactation biology by placing it within the context of variation observed among mammals.

Life History and Lactation Biology

Table 3.1 defines some key terms used to describe aspects of lactation biology in humans and other mammals. The comparative methods of zoology data can be used to distinguish the evolutionarily derived features of human life history and lactation biology from those that are shared with other mammals. Considered in the light of detailed physiological and epidemiological data available on the underlying mechanisms and associated health outcomes of variation in human lactation, identification of the evolutionarily derived characteristics of human infant and young child feeding

Table 3.1. Terms used to describe aspects of lactation biology.

Plesiomorphies	shared primitive characters no older than the last common ancestor of a phylogenetic group of organisms
Synapomorphies	evolutionarily derived or specialized characters shared only by one phylogenetic group of organisms
Apomorphies	derived characters that are not shared with an organism's ancestors and therefore likely to be specialized and recent adaptations
Lactation	the ability to secrete immunologically active and nutritious milk from ventral epidermal glands
Suckling	the mechanical process of extracting milk from the mammary glands and/or ducts. In humans, suckling is a component of "nursing" or "breastfeeding" by a mother or baby. It is interesting that in English and many other languages such verbs carry both active and passive meanings.
Exclusive suckling	a life history phase during which a juvenile mammal derives all nutrients from maternal milk. It is often referred to as "infancy" in nonhuman mammals.
Transitional feeding	a life history phase during which nutrition is derived from a combination of maternal milk and other foods foraged by the infant, its parents, or others. It is poorly described for most primates.
Complementary feeding	feeding of nutritionally rich and relatively sterile combinations of foods acquired and processed by care givers to breast-fed infants and toddlers after about six months of age
Weaning	the termination of suckling
Weanling	a life history phase during which a recently weaned juvenile mammal must forage for itself and subsist on foods similar or identical to those selected by adults
Infancy	in humans, the period between birth and the first birthday; in other mammals, the period of dependency on mother's milk
Young childhood	in humans, the period between the first and third birthday
Preschoolers	in humans, children between the ages of one and five years
Family foods	raw foods and combinations of foods collected, processed, and shared by older juveniles and adults and consumed by older members of the family

should allow us to construct a model for an evolved, optimal pattern of human infant and young child feeding (IYCF) practices (Sellen 2006, 2007, 2009).

Mammals vary in age at weaning, as well as in many other characteristics that together describe their *life history*, i.e., the temporal organization over the lifetime of individuals of major biological events (e.g., age at first reproduction, achievement of adult body size, or death) and phases (e.g., length of time between conception and birth or between births; Horn 1978). Evolutionary theory suggests life history variation is an adaptive response to natural selection within physiological, ecological, and social constraints (Futuyma 1998; Stearns 1992). Life history theory aims to explain why life cycles vary among species in terms of the Darwinian mechanisms that link more or less species-typical patterns of growth and development, demography, sociality, and ecology (Daan and Tinbergen 1997; Leigh and Blomquist 2007). Evolutionary anthropologists agree that a suite of complex biocultural adaptations have emerged during the coevolution of human diet and life history and continue to generate and test new hypotheses for the underlying causal mechanisms (Hawkes 2006a; Hill and Kaplan 1999; Mace 2000).

Lactation is a defining characteristic of mammals (Hartmann, Morgan, and Arthur 1986; Plaut and Carraway 2002). All surviving mammals retain lactation as a key adaptation that contributes to the organization of life history characteristics through four basic functions present as plesiomorphies: (1) transfer protective functions of fully developed immune system across generations (Goldman 2007); (2) optimize litter size to allow efficient allocation of maternal investment across sib sets (Bronson 1985); (3) facilitate efficient reproduction in unpredictable environments lacking special foods for young (Dall and Boyd 2004; Hartmann et al. 1984); and (4) increase behavioral flexibility and opportunities for learning (Peaker 2002). To support these functions, many biochemical, physiological, and developmental pathways involved in lactogenesis, mammary development, immunological activity, milk transport proteins, and metabolic adaptation during lactation appear to be similar in many mammals.

In spite of the challenges inherent in making inferences about a process that depends largely on soft tissues unlikely to be preserved as fossils, beginning with Darwin (1872) a number of scenarios have been proposed to explain the evolution of mammary glands (Oftedal 2002a; Table 3.2). Careful comparative studies indicate that lactation probably evolved before placentation and hair between 210 and 190 MYA (Cifelli et al. 1996; Hartmann et al. 1984; Kumar and Hedges 1998; Luckett 1993; Luo, Crompton, and Sun 2002; Messer et al. 1998; Oftedal 2002a), initially as an adaptation to keep water-permeable eggs moist (Oftedal 2002b), later as an adaptation to transfer immune factors to offspring (Blackburn 1993; Hayssen and Blackburn 1985), and still later to make efficient use of maternal body fat and other stored nutrients in feeding offspring and spacing births (Pond 1997). Species-specific characteristics of lactation biology vary greatly in relation to life history and reflect phylogenetic differences in the selective response to shifts in disease ecology, foraging

Table 3.2. Relationships between life history and lactation biology. (References in Sellen 2006)

(i) Milk concentration decreases with maternal and neonatal size.
(ii) Milk fat and protein are positively correlated, and both negatively correlated with sugar.
(iii) Milk energy output at peak lactation scales with basal metabolic rate according to Kleibers Law.
(iv) Marsupials commonly overlap lactation with gestation of younger offspring whereas most placental species do not.
(v) The period between first solid food consumption and weaning is long in species with single, precocial young, and provisioning may occur, whereas first solid food is usually eaten near weaning in polytokous species with altricial young.

strategy, and constraints on growth and development. Table 3.2 summarizes some key trends linking variation in lactation biology and life history across mammals.

Shared, Derived Features of Lactation Biology in Nonhuman Primates

More is known about the range of life history variation observed among nonhuman primates and hominids (Leigh and Blomquist 2007; Pereira and Fairbanks 2002; van Schaik et al. 2006) than about variation in primate lactation biology (Sellen 2006; Wright 1990). Recent work suggests the common ancestor of primates weighed between 1 and 15 g and therefore had high metabolic, reproductive, and predation rates, and that body size remained below 50 g during the early Eocene primate radiations (Gebo 2004). Extant primates range in size by an order of magnitude (Harvey and Clutton-Brock 1985) and are characterized by comparatively slow development and postnatal growth rates and greater longevity (Charnov and Berrigan 1993; Harvey and Clutton-Brock 1985; Harvey, Martin, and Clutton-Brock 1987). The few available data on variation in primate lactation biology suggest all species share common adaptations to meet infant nutritional needs conditioned by this characteristic slow life history.

Milk Composition

Primates are unusual among mammals in that the milk they produce is lower in volume; more dilute; lower in energy, fat, and protein; and higher in lactose than would be predicted by body size (Martin 1984; Peaker 2002). Length of lactation is relatively long and exceeds that of gestation (Hartmann et al. 1984). The gross composition of milk does not vary widely across nonhuman primate species with differences in body size, reproductive rates, patterns of maternal care, or other life history characteristics (Buss 1971; Dufour and Sauther 2002; Fomon 1986; Hinde 2006; Kanazawa et al.

1991; Patino and Borda 1997; Power, Oftedal, and Tardif 2002). It has long been hypothesized that these shared characteristics of primate milk coevolved with low reproductive rates and slow life histories relative to body size (Ben Shaul 1962; Lee and Bowman 1995; Oftedal 1984a; Peaker 2002). Thus, a lower protein concentration of primate milks coevolved with slower growth rates (Oftedal 1984b; Oftedal 1986); lower fat concentration coevolved with the behavioral ecology of continuous infant carrying (which facilitates frequent suckling and is unusual in any other order of mammals) (Ben Shaul 1962; Martin 1984; Oftedal and Iverson 1987; Tilden and Oftedal 1997); and a relatively high lactose content coevolved with the lower fat storage in adult females and low fat content of milk. This composition may also be linked to faster rates of postnatal brain growth, although there is no evidence that levels of long chain polyunsaturated fatty acids (LCPUFAs) increase among primates with rates of postnatal brain growth (Milligan 2008 in press; Robson 2004).

Juvenile Feeding Ecology

One correlate of slow development is a relatively early maturation of the gastrointestinal tract in primates. There is little clustering of gut maturational changes around the species-typical age at birth and weaning (Sangild 2006), and primates are therefore able to begin consuming milk even if born preterm. In contrast to many other mammals, they are generally viable from about 70 percent of the average length of gestation. Figure 3.1 illustrates how, from a nutritional perspective, primate postnatal life can be divided for all species into three phases (exclusive suckling, transitional feeding, and weaning) separated by two key life history variables (age at first solid food and age at weaning) (Sellen 2007, 2009). Juvenile daily intake of energy and specific nutrients increases from birth and is entirely due to greater milk intake during exclusive suckling. After weaning, a further increase in total intake occurs by means of independent foraging.

Few data are available to assess the length of transitional feeding in primates or the relative nutritional contribution of milk versus foraged foods (Sellen 2006). Variation in transitional feeding may be substantial both within and between species but has yet to be fully described and explained (Lee 1996, 1999). Data are insufficient to test a hypothesis that relative length of transitional feeding is inversely related to adult diet quality (Sellen 2009). The transition to weaning is a gradual process in orangutans (van Noordwijk and van Schaik 2005) and possibly chimpanzees (Nishida, Ohigashi, and Koshimizu 2000). There is limited evidence that parental provisioning occurs in apes (Nishida and Turner 1996). However, in the absence of observational data, it is often assumed that infants in most nonhuman primate infants wean relatively abruptly and begin to forage on foods similar to those selected by the mother, processing them largely for themselves, and that provisioning of juveniles is rare or absent (Galdikas and Wood 1990; Knott 2001).

Data from captive populations indicate that nonhuman primate weaning age is scaled to other life history traits such as gestation length, birth weight, and adult

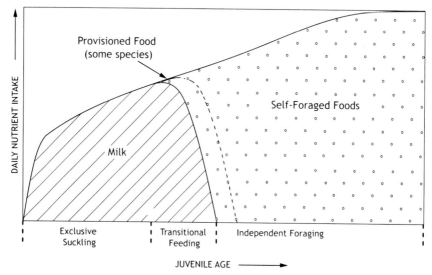

Figure 3.1. Juvenile feeding ecology in nonhuman primates. Modified after Sellen 2007.

weight and developmental events such as age at molar eruption (Charnov and Berrigan 1993; Harvey and Clutton-Brock 1985; Lawrence 1989: 245; Lee, Majluf, and Gordon 1991; Robson, van Schaik, and Hawkes 2006; Smith 1991; Smith 1992). However, such traits do not reliably predict age at weaning for all primate species, suggesting that age at weaning is labile. Last suckling is very difficult to observe directly in the wild, but studies suggest that weaning age is plastic in most wild populations and sensitive to ecological factors that constrain maternal ability to meet the increasing energy needs of growing offspring and the ability of infants to survive without mother's milk (Altmann 1980; Altmann and Alberts 2003; Lee 1996, 1999).

Infant Dietary Requirements

Few data are available on age-specific changes in energy or nutrient requirements of nonhuman primates, of the total energy or nutrient costs of growth and maintenance during infancy, or of the proportion of energy and nutrient needs met by milk or other foods consumed. Average infant energy intakes of several captive larger monkey species (such as baboons) estimated from observation of free-feeding fall in the range 0.837–1.255 MJ/kg/d (National Academy of Sciences 1989; Nicolosi and Hunt 1979). Average requirements or usual intakes in the wild are likely to differ. For example, a study of free-living yearling baboons (*Papio cynocephalus*) estimated minimum total energy requirements for growth and maintenance at only 0.383 MJ/kg/d (Altmann 1998). Since it is difficult to identify studies that estimate the concentration of key nutrients such as vitamin A, vitamin D, iodine, and calcium (Power,

Oftedal, and Tardif 2002), few conclusions can be drawn for any species about how milk nutrient content varies with maternal diet or about the extent to which exclusive suckling and milk consumption during transitional feeding satisfy age-specific nutrient requirements. Although evolved associations between feeding ecology and milk composition might be predicted across species, the data are too scant to test nutritional ecological hypotheses.

Maternal Reproductive Ecology

Little is known about the extent to which nonhuman primates share a capacity for maternal accommodation of lactation performance in response to moderate decreases in maternal energy or nutrient intakes. Field observations of several species indicate that lactation places a significant metabolic demand on mothers and that limited mechanisms exist to accommodate the costs of protecting their infants against fluctuations in milk volume and composition when conditions are adverse (Altmann and Samuels 1992; Dufour and Sauther 2002; Lee 1987). Free-living yearling baboons are estimated to consume 2.251 MJ/d, of which approximately 40 percent (0.900 MJ/d) comes from milk, suggesting their mothers bear the cost of minimal energy requirement cited above (Altmann 1998). No studies have shown conclusively that lactating nonhuman primate mothers are able to reduce the daily costs of milk production using fat stored during pregnancy. Indirect evidence from several captive studies suggests energetic costs of lactation are accommodated by energy-sparing physiological adaptations (Roberts, Cole, and Coward 1985), reductions in physical activity (Bercovitch 1987), and shared care of infants (Tardif, Harrison, and Simek 1993). Wild lactating females increase their intake of high energy foods (Boinski 1988), overall food energy (Sauther 1994), and time allocated to foraging (Altmann 1980; Dunbar and Dunbar 1988), particularly when forage quality is poor (Dunbar, Hannah-Stewart, and Dunbar 2002). Among wild great apes, female reproductive biology seems designed to avoid conception under food stress rather than to protect mothers from nutritional deficiency during lactation (Bentley 1999; Nishida et al. 2003). Field observations indicate food availability is so unpredictable that conception cannot be timed so that birth will occur during periods of highest food availability and is therefore most likely to occur during periods of positive maternal energy balance (Knott 2001).

Shared, Derived Features of Human Lactation Biology

Much evidence suggests that the four basic functions of lactation in the infant and mother, the basic composition of human milk, and its mechanism of secretion and delivery have remained unchanged during 7 million years of human evolution. We can infer that such features must have been present in our last common ancestor with apes (which lived approximately 6–7 MYA) and in all subsequent hominid species including those ancestral to humans (i.e., various members of the genera *Ardipithecus, Australopithecus,* and *Homo*). This is striking given that during this period there occurred a shift to bipedal locomotion, radical dental and cranial adaptations to a more omnivorous diet, a large increase in brain size, a doubling of adult body size, an even larger increase in the length of the juvenile period and of total lifespan, a shortening of birth intervals and an increase in female post-reproductive lifespan. In fact, the unique features of human lactation appear to be linked primarily to shifts in the behavioral ecology of juvenile feeding behavior and the regulation of maternal lactation performance rather than changes in the physiology of milk secretion per se.

Uniquely Derived Features of Human Life History and Lactation Biology

There has been considerable debate about whether and why human life history differs from the typical primate pattern (Blurton Jones, Hawkes, and O'Connell 1999; Hawkes and Paine 2006; Hill and Kaplan 1999; Kaplan et al. 2000; Walker et al. 2006). A consensus has recently emerged that, in comparison to other primates, humans have evolved four distinctive life history traits: slow maturation, long life spans with slow aging, postmenopausal longevity, and weaning before independent feeding (Hawkes 2006a, Robson et al 2006). Table 3.3 summarizes some key differences in life history parameters linked to lactation among our closest relatives using values obtained for female wild great apes and human hunter-gatherers with natural fertility. In addition, a number of features of human lactation biology appear to have been retained as plesiomorphic with mammals and synapomorphic with nonhuman primates (Sellen 2006). Table 3.4 summarizes the shared and derived characteristics of human lactation biology.

Although not the largest extant ape, humans have the slowest life history as evidenced by a markedly later age at maturity or first birth, a longer period of nutritionally "independent" growth between weaning and maturity, longer maximum lifespan, and longer potential adult lifespan. Not all aspects of human life history are slowed, however. Gestation lengths are similar for all living ape species despite appreciable variation in size at maturity. Healthy human neonates are relatively large for gestational age and relative to maternal body size, indicating faster fetal growth rates. The estimates of human weaning age and relative weaning weight fall at the lower end of the great ape range, and infants in hunter-gatherer populations are weaned after relatively smaller postnatal weight gain than are ape infants (Hill and Hurtado 1996). Most striking, the human birth interval is exceptionally short, both in absolute time and relative to body size. Although birth intervals rarely exceed four years in natural fertility human populations (Wood 1990), half of all randomly selected closed birth intervals exceed four, five, and eight years in wild gorillas, chimpanzees, and orangutans, respectively (Galdikas and Wood 1990). Since fertility ends at similar ages in human and chimpanzee females, the "species-typical" rate of human reproduction is higher (Blurton Jones, Hawkes, and O'Connell 1999).

Current international recommendations based on clinical and epidemiological data provide a compelling model for the evolved pattern of human IYCF practices

Table 3.3. Phylogenetic relationships of great ape species and average values for selected life history parameters.

	Estimate time of divergence from hominid lineage, MYA	Adult female weight (range), kg	Neonate weight / maternal weight, %	Weaning weight / maternal weight, %	Age at weaning, years	Period of independent growth, years	Age at first birth, years	Gestation length, years	Birth interval, years	Maximum lifespan, years	Potential adult lifespan, years
Homo sapiens	—	47.0 (38–56)	5.9	0.2	2.8	16.7	19.5	0.7	3.7	85.0	65.5
Pan troglodytes	5–7	35.0 (25–45)	5.4	0.3	4.5	8.8	13.3	0.6	5.5	53.4	40.1
Pan paniscus	5–7	33.0 (27–39)	4.2	—	—	—	14.2	0.7	6.3	50.0	35.8
Gorilla gorilla	6–8	84.5 (71–98)	2.3	0.2	2.8	7.2	10.0	0.7	4.4	54.0	44.0
Pongo pygmaeus and *Pongo abelii*	12–15	36.0	4.3	0.28	7.0	8.6	15.6	0.7	8.1	58.7	43.1

Source: Robson, van Schaik, and Hawkes 2006, and articles cited therein

Table 3.4. Evolutionary classification of characteristics of human lactation biology.

	PLESIOMORPHIC Shared with all mammals	SYMPLESIOMORPHIC Shared with most primates	APOMORPHIC Unique to humans
Postnatal immune defense	X		
Optimal postnatal nutrition	X		
Fertility regulation	X		
Developmental window for learning	X		
A period of exclusive lactation yields optimal benefits to mothers and offspring	X		
Low protein, fat and high lactose milk content		X	
Slow infant growth		X	
A period of transitional feeding yields optimal benefits to mothers and offspring		?	
Age at weaning highly labile relative to other life history traits		?	
Complementary feeding			X
Increased plasticity in length of lactation relative to body size			X
Reduced infant energy needs			?
Significant buffering of lactation by fat storage in pregnancy			?

because they are predictive of optimal growth and development of healthy newborn humans in all favorable environments (Garza 2006; Sellen 2006; World Health Organization and Fund 2003). Figure 3.2 schematizes the evolved template for human IYCF, which includes (1) initiation of breast feeding within an hour of birth, (2) a six-month period of exclusive breast feeding, (3) introduction of nutrient-rich and pathogen-poor complementary foods at about six months of infant age, (4) continued breastfeeding through the second year, and (5) a package of "responsive care-giving." Comparison with Figure 3.1 suggests that the derived features of human lactation biology include (1) complementary feeding and (2) early and flexible weaning. Note that in figure 3.2 the period of "transitional feeding" is labeled as "complementary feeding" to highlight the important difference between provisioned food in nonhuman primates (which are rare and always raw) and complementary food in humans, which is necessary for optimal health and usually cooked.

Complementary Feeding

The pattern of transitional feeding appears to be fundamentally different in humans. The most remarkable change is the use of complementary foods, which is unique among mammals (Knott 2001). Although it is not known whether nonhuman primate infants have a similar biological capacity, a wealth of data on the trajectory of infant development of feeding competency and changes in the nutritional needs of growing infants in relation to maternal milk supply supports the hypothesis that humans evolved to begin consuming complementary foods at approximately six months

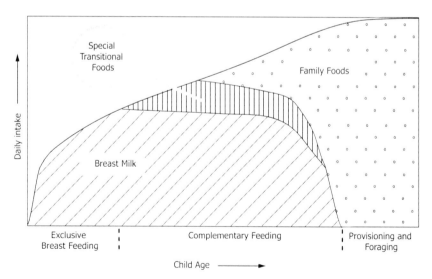

Figure 3.2. Evolved template for human infant feeding. Modified after Brown and Dewey 1998.

of age. Overwhelming clinical and epidemiological evidence demonstrates that infants have not evolved to make efficient use of other foods before six months (Dewey et al. 2001; Kramer et al. 2003) and may suffer deficits and increased morbidity if not exclusively breastfed (Black et al. 2008; Kramer and Kakuma 2002). After approximately six months of age complementary and family foods increasingly contribute to the diet as chewing, tasting, and digestive competencies develop. Frequency of suckling and volume of milk consumed do not necessarily diminish after six months in healthy babies, and breast milk remains an important, sterile source of nutrients and immune protection during a complementary feeding phase that can continue at least until the third year of life. Published ethnographic reports indicate that the duration of exclusive breastfeeding is extremely variable in nonindustrialized small-scale societies. One study estimates the modal age of introduction of liquid and solid foods in a sample was approximately four and six months, suggesting that the age-related pattern for introduction of complementary foods concords closely with the current clinical recommendations for normal, healthy children (Sellen 2001b) and that a sizeable proportion of infants in these populations may have been exclusively breast fed for six months (Sellen 2001a).

Early and Flexible Weaning

The degree of flexibility in age at weaning is unusual among primates and probably a distinctive, derived characteristic. Humans are the only primates that wean juveniles before they can forage independently (Hawkes 2006b). The targeting and sharing of high yield, nutrient dense foods that entail high acquisition and processing costs and use of heat treatments and combination of raw foods in "cuisine" are human specializations (Bird 2001; Conklin-Brittain, Wrangham, and Smith 2002). We are also unusual in the extent to which human mothers recruit additional help with young child feeding and care (Blurton Jones 1993; Trevathan and McKenna 1994). Thus, weaning marks a shift to food sharing, not feeding independence.

It is difficult to conclude that humans have evolved a species-specific, global optimum weaning age. Current international recommendations are based on evidence that infants benefit from breastfeeding into the third year (World Health Organization and Fund 2003) but the clinical data are silent on whether there is any upper age limit at which breastfeeding ceases to be of some benefit to children (Fewtrell 2003; Marquis and Habicht 2000). Continued breastfeeding must have remained a strongly selected component of ancestral maternal strategies because of its powerful anti-infective properties (Heinig 2001; Labbok, Clark, and Goldman 2005; McGuire 2005) and nutritional benefits to infants (Black et al. 2008; Jones et al. 2003). Breastfeeding is also associated with benefits to mothers, including healthy maintenance of energy and nutrient balance and reduced risk of disease (Labbok 2001).

The enormous diversity in human breastfeeding and complementary feeding patterns has long been a focus of lactation researchers (Hartmann et al. 1984; Sellen

2002) and anthropologists (Sellen and Smay 2001). Ethnographic evidence from recent and contemporary foraging populations indicates that human weaning age is extremely variable within and between groups and that the process of weaning could be gradual (Fouts and Lamb 2005; Konner 2005) or (less commonly) abrupt (Ford 1945; Fouts, Hewlett, and Lamb 2005). If initiated, the duration of human lactation in recent and contemporary societies ranges from a few hours to more than five years for some individuals, spanning almost the entire range observed for other mammals (Hartmann and Arthur 1986; Kennedy 2005; Sugarman and Kendall-Tackett 1995). Humans also wean over a wide range of infant sizes.

Data rigorously collected from hunter-gatherer populations suggest a mean weaning age of 2.8 years (Alvarez 2000). If this is an indication of an evolved, species-typical value, it is well below the five to seven years predicted for primates with our life history parameters (Dettwyler 2004). Indeed, a mean weaning age of four years reported among the !Kung is the largest reliable estimate for a foraging population (Konner 1977). It is estimated that breastfeeding beyond two years was the norm in between 75 percent (Nelson et al. 2005) and 83 percent (Barry III and Paxson 1971) of small-scale societies and that the modal age at weaning was approximately thirty months (Sellen 2001a, 2001b). Thus, a sizeable proportion of infants in these populations may have been partially breast fed for more than two years. Isotopic studies also suggest that children in past populations often shifted to solid foods before two years of age while continuing to breastfeed (Herring, Saunders, and Katzenberg 1998; Schwarcz and Wright 1998).

Taken together, such indicators suggest that the age-related pattern for introduction of complementary foods and termination of breastfeeding observed in small-scale societies was closer to the current clinical recommendations for normal, healthy children than it is in many urban or rural communities today. Nevertheless, evidence accumulates that the structure of women's work is an important factor constraining breastfeeding in all types of society (Nerlove 1974; Sellen 2002). Observation in contemporary human societies shows lactation behavior is sensitive to maternal workload and the availability of cooperative childcare and feeding (Baumslag and Michels 1995; Sellen 2001b). Weaning age is later among foragers than among subsistence herders and farmers (Sellen and Smay 2001), among whom women often do more kinds of work that separates them from their infants for extended periods.

Infant Dietary Requirements

Cumulative energy requirement for male babies average 374.2 MJ between birth and six months and 959.4 MJ between birth and twelve months (recalculated from data in Butte, Henry, and Torun 1996). Depending on age and sex, estimated mean total energy requirements for growth and maintenance in the first year range between 0.351 and 0.372 MJ/kg/d (1.347 and 3.519 MJ/d). These estimates fall below the estimates for free-living yearling baboons (Altmann 1998) and well below those for captive large-bodied cercopithecines (National Academy of Sciences 1989; Nicolosi and Hunt 1979). These energy requirements are regarded as universally valid because healthy infants from different geographic areas show relative uniformity of growth, behavior, and physical activity (Butte 1996; Garza 2006). Thus, although comparative data are scant, human infants appear to have low energy requirements due to comparatively slower growth in comparison to other primates.

Maternal Reproductive Ecology

Human peak milk volume corresponds to a mean infant energy intake of 2.87 MJ/d (Prentice et al. 1996), which is well in excess of healthy infant requirements in the first six months. The crude energetic cost of human lactation has been estimated from measurements of daily milk intake among predominantly breastfed infants observed for two years postpartum and is estimated to be ~1,686 MJ, more than half of which is borne in the first year of infant life (recalculated from data in Prentice et al. 1996). This corresponds to a mean daily additional cost of approximately 2.3 MJ/d (actually 2.7 MJ/d in the first six months) (Prentice et al. 1996). Thus, the daily cost of lactation is potentially high (~25–30%) in relation to average total energy expenditure for a moderately active (1.7 × resting metabolic rate) nonpregnant, nonlactating woman weighing 50 kg (calculated from equations in Food and Nutrition Board 1989). However, two mechanisms allow mothers to accommodate the cost of lactation, both of which appear to be derived for humans relative to our nonhuman primate ancestors.

First, depletion of the maternal fat laid down before and during pregnancy has the potential to subsidize lactation by ~118.6 MJ (0.325 MJ/d) in the first year. Fat storage demands the largest proportion (~71%) of additional energy needed to sustain a healthy pregnancy in nonchronically energy deficient women (Durnin 1987; Lawrence et al. 1987; Prentice et al. 1996). Nevertheless, reductions in basal metabolic rates and physical activity (Durnin et al. 1985; Prentice et al. 1996) ensure that for many women average daily costs of pregnancy (~0.7 MJ/d) are low (~8%) in relation to the usual dietary energy intakes and requirements of healthy nonpregnant, nonlactating women (~8.78 MJ/d). In favorable conditions, the average woman begins lactation with approximately 125 MJ of additional fat accumulated during pregnancy. Second, feeding of nursing infants using safe and nutritionally adequate complementary foods from 6 months of age can result in maternal energy savings of almost 1.8 MJ/d in the first year.

Together, healthy fat depletion and complementary feeding reduce the actual cost of lactation estimated to satisfy infant and young child needs for two years by 1,023.6 MJ, or almost 61 percent. On a daily basis this reduces the net additional costs from ~2.3 MJ/d to ~0.9 MJ/d. For many women, this represents between 10 and 20 percent of usual total energy expenditure. Healthy people unconstrained in their access to food or choice of activities can comfortably increase energy intake, decrease physical activity, or both to accommodate increases in daily energy requirement of up to 30 percent. Despite these adaptations, however, the average daily energetic cost of human lactation is potentially higher than that of pregnancy (~2.3 MJ/d versus ~0.7 MJ/d),

One corollary is that human lactation performance is well buffered from fluctuations in maternal condition and nutrient supply (Allen 1994; Institute of Medicine (US) Subcommittee on Nutrition during Lactation 1991a, 1991b; Prentice 1986; Rasmussen 1992). Aerobic exercise and gradual weight loss have no adverse impact on milk volume or composition, infant milk intake, infant growth, or other metabolic parameters (Dewey et al. 1994; Lovelady, Lönnerdal, and Dewey 1990; McCrory 2000, 2001). A single intervention study has suggested that milk production can be improved by maternal food supplementation during lactation (Gonzalez-Cossio et al. 1998). However, most studies suggest lactation is rarely compromised even when mothers are marginally undernourished, engaged in high levels of physical activity, have experienced multiple pregnancies, or lose weight and fat with age and by season, (Adair, Pollitt, and Mueller 1983; Brown et al. 1986; Lunn 1985; Prentice et al. 1986; Prentice et al. 1981; Winkvist et al. 1994).

A Model of the Evolved Pattern of Human Infant and Young Child Feeding

The conceptual framework developed above and schematized in Figure 3.3 highlights similarities and differences in human and nonhuman primate lactation biology. The

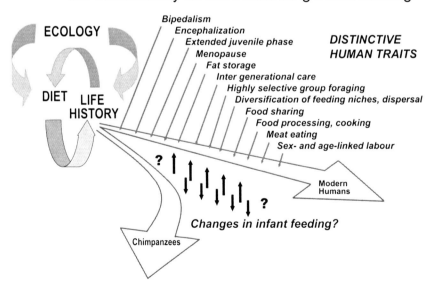

Figure 3.3. Co-evolutionary forces influencing infant feeding among human ancestors. Modified after Sellen and Kingston, unpublished MS.

pattern of similarities and differences indicates that the ancestors of modern humans evolved an unusually flexible strategy for feeding young. In particular, all available data are consistent with a hypothesis that complementary feeding evolved as a facultative strategy that provided a unique adaptation for resolving tradeoffs between maternal costs of lactation and risk of poor infant outcomes. A flexible lactation strategy was likely adaptive in the environments in which humans evolved, and its behavioral and physiological components have been retained. However, such flexibility has created potential for human breastfeeding and complementary feeding practices in many contemporary populations to track outside a range that was naturally selected to optimize healthy survival of mothers and their offspring. Indeed, epidemiological studies show that significant mismatch between the evolved pattern of infant and young child feeding and actual feeding is harmful in many settings (Bahl et al. 2005; Coutsoudis 2005; Edmond et al. 2006; Black et al. 2008).

Human Demography

As a species humans are particularly good at keeping young alive in a wide range of habitats. Infant and weanling survival is much greater among foragers (60–70 percent; Hewlett 1991; Lamb and Hewlett 2005) than among apes (25–50 percent; Hill and Hurtado 1995; Hill and Kaplan 1999; Kennedy 2005), and greater still in nonindustrial herding and farming economies (70–90 percent; Hewlett 1991, 2005; Kaplan et al. 2001; Nishida et al. 2003; Sellen and Mace 1999). The human population now exceeds 6 billion while total populations of great ape species likely remained in the low thousands even without their recent decimation by human habitat destruction and hunting (Knott 2001). Thus, an early age at weaning suggests that ancestral humans evolved an unusual capacity to reduce the length of exclusive and transitional feeding without increasing infant mortality. Nevertheless, an inverse relationship between birth interval and child survival is mediated by breastfeeding (Black et al. 2008) and birth intervals below two years are risky for older siblings.

Shortened birth interval is currently regarded by anthropologists as one of the most evolutionarily significant human deviations from the expected pattern of great ape life history (Galdikas and Wood 1990; Kennedy 2005; Knott 2001; Robson, van Schaik, and Hawkes 2006; Sellen 2006). Among ancestral mothers, shortening of the periods of exclusive lactation or transitional feeding, or both, likely reduced birth intervals (by accelerating the return of ovarian cycling) and may have improved subsequent birth outcomes (by reducing maternal depletion). Shortened birth spacing would have increased maternal fitness only if it did not increase offspring mortality. Reduced juvenile mortality could be achieved if the nutritional and antibiotic components of breast milk were substituted by other kinds of foods, or if infant development was accelerated so that the period of nutritional dependency was shortened (Foley 1995; Hammel 1996; Pennington 1996).

Relevance to Contemporary Public Health Challenges

Despite burgeoning biomedical (Goldman 2001), and anthropological (van Esterik 2002) research on human lactation in recent decades, few scholars have asked broader evolutionary questions about which characteristics of human lactation biology reflect evolutionarily conserved design features and which aspects, if any, reflect a distinctively human phenotype (Wray 1991). An evolutionary perspective provides insight into why contemporary patterns of IYCF often deviate from the optimal pattern indicated by clinical and epidemiological evidence. Human mothers are physiologically and behaviorally adapted to exercise more choice in the patterns and duration of full and partial breastfeeding than do other primates. The evolution of the use of complementary foods to facilitate physiologically appropriate early weaning relative to other species has created potential for physiologically inappropriate early weaning and introduction of other foods to breastfed infants. Mismatch between optimal and actual infant feeding practices in contemporary populations is widespread and presents a major public health challenge (Huffman and Martin 1994; Quandt 1985; Underwood and Hofvander 1982). Common practices such as discarding of colostrum (Gunnlaugsson and Einarsdottir 1993; Rizvi 1993), use of supplementary foods before breastfeeding begins (Akuse and Obinya 2002; Pérez-Escamilla et al. 1996), and reduced breast milk intake due to early introduction of formula and other substances and early weaning (Heinig et al. 1993) are all associated with infant illness and death.

Remaining Questions about the Evolution of Human Lactation and Complementary Feeding

This discussion has drawn on the rich set of data now available on human lactation biology and is limited by the paucity of available data on nonhuman lactation (Hinde 2007; Sellen 2009). A full understanding of the significance of adaptations for infant and young child feeding in the coevolution of human diet and life history will require the collection of a richer comparative database on nonhuman primate lactation biology and the resolution of number of outstanding questions (Sellen 2007). Primary among these are: (1) the extent of behavioral and physiological variation in scheduling and dietary impact of transitional feeding in nonhuman primates, and its socioecological correlates (such as quality of adult diet); (2) the absolute timing of steps in the evolution of complementary feeding, and their relation to other key derived human characteristics (such as the origins and maintenance of bipedal locomotion, increased brain size, reduced sexual dimorphism, tool use, food sharing, cooking, alloparenting); (3) how humans evolved to meet the physiological and psychosocial challenges to the successful establishment of exclusive breastfeeding that are observed in contemporary populations; (4) how humans evolved to meet the complex challenge of introducing appropriate complementary foods in a timely manner while satisfying other demands on caregivers; and (5) how human lactation biology evolved to protect low birth weight and premature infants, and to reduce the risks of maternal-to-child transmission of pathogens.

Summary

Unlike other species, the life histories of mammals have coevolved with the special adaptive advantages and physiological demands of lactation biology. This review has suggested that human patterns of lactation and complementary feeding are intimately linked with the evolution of a distinctive set of human life history variables. Complementary feeding and fat storage in pregnancy probably evolved in the last 5–7 million years as unique and important human adaptations that together reduce the energetic and opportunity costs of lactation for mothers and the potential fitness costs of relatively early transitional feeding and weaning for infants and young children. The human lactation span is comparatively short. The lower boundary for safe complementary feeding has been set at around six months for normal term and preterm babies by constraints on the evolution of physiological factors such as the growth and maturation of infant systems affecting immune, feeding, and digestive competency. The upper bound for safe weaning has been set above two years of infant age by similar constraints, but the evolution of complementary feeding has introduced enormous behavioral flexibility in maternal response to social and ecological constraints.

This evolutionary perspective provides insight into why, in today's world, young child feeding practices are clinically sub-optimal for most children and their mothers, and why many people in both rich and poor societies fail to adopt recommended feeding practices. Understanding the ultimate evolutionary causes of human variability in young child feeding can provide insights on the proximate causes of patterns of breastfeeding and complementary feeding that subsequently lead to poor health outcomes for mothers and babies. Such insight may help in the design of interventions to promote improved infant feeding practices.

Conclusion

In this chapter I have summarized recent conclusions about the unique characteristics of human life history, discussed how they may be related to unique characteristics of human lactation biology, and briefly reviewed data on variation in lactation patterns among nonhuman patterns of IYCF among pre-industrial human societies and ancient populations. I have highlighted two observations about the patterns of infant and young child feeding in contemporary human societies that are puzzling. First, the proportion of newborns that breastfeed exclusively for six months, receive timely and appropriate complementary foods, and continue to breastfeed into their third year is small, even though overwhelming evidence suggests such a pattern is optimal for most healthy, term infants (Black et al. 2008). Second, humans tend to wean their babies significantly earlier than most other apes do, even though children depend on care givers for subsistence much longer than do the offspring of any other mammal (Paine and Hawkes 2006).

A review of zoological, anthropological, and nutritional data suggests that these apparently paradoxical observations are evolutionarily linked. It provides an evolutionary perspective on why optimal infant and young child feeding is so rare and difficult to promote in modern human societies that are far removed from the original conditions shaping human adaptation. Thus, contemporary challenges in maternal and child nutrition can be understood in terms of the evolutionary biology of human lactation and complementary feeding. Given our evolutionary history, we would predict that contemporary parents show a marked tendency to shorten the periods of exclusive and any breastfeeding.

Acknowledgements

I am grateful to John Kingston for his conceptual and technical contributions to Figure 3.3. This work was made possible by generous financial support from the Canada Research Chairs Program (CRC), Canadian Institutes of Health Research (CIHR), and the Canadian Foundation for Innovation (CFI).

References

Adair, L. S., E. Pollitt, and W. H. Mueller. 1983. Maternal anthropometric changes during pregnancy and lactation in a rural Taiwanese population. *Human Biology* 55: 771–87.

Akuse, R. M., and E. A. Obinya. 2002. Why healthcare workers give prelacteal feeds. *European Journal of Clinical Nutrition* 56: 729–34.

Allen, L. H. 1994. Maternal micronutrient malnutrition: Effects on breast milk and infant nutrition, and priorities for intervention. *SCN News* 11: 21–24.

Altmann, J. 1980. *Baboon Mothers and Infants.* Cambridge, MA: Harvard University Press.

Altmann, J., and S. C. Alberts. 2003. Variability in reproductive success viewed from a life-history perspective in baboons. *American Journal of Human Biology* 14: 401–9.

Altmann, J., and A. Samuels. 1992. Costs of maternal care: Infant carrying in baboons. *Behavioral Ecology and Sociobiology* 29: 391–98.

Altmann, S. A. 1998. *Foraging for Survival: Yearling Baboons in Africa.* Chicago: Chicago University Press.

Alvarez, H. P. 2000. Grandmother hypothesis and primate life histories. *American Journal of Physical Anthropology* 113: 435–50.

Bahl, R., C. Frost, B. R. Kirkwood, K. Edmond et al. 2005. Infant feeding patterns and risks of death and hospitalization in the first half of infancy: Multicentre cohort study. *Bulletin of the World Health Organization* 83: 418–26.

Barry III, H., and L. M. Paxson. 1971. Infancy and early childhood: Cross-cultural codes 2. *Ethnology* 10: 466–508.

Baumslag, N., and D. L. Michels. 1995. *Milk, Money, and Madness: The Culture and Politics of Breastfeeding*. Westport, CT: Bergin & Garvey.

Ben Shaul, D. M. 1962. The composition of the milk of wild animals. *International Zoological Yearbook* 4: 333–42.

Bentley, G. R. 1999. Aping our ancestors: Comparative aspects of reproductive ecology. *Evolutionary Anthropology* 7: 175–85.

Bercovitch, F. B. 1987. Female weight and reproductive condition in a population of olive baboons (*Papio anubis*). *American Journal of Primatology* 12: 189–95.

Bird, D. W. 2001. Human foraging strategies: Human diet and food practices. In *Encyclopedia of Evolution*, ed. M. Pagel, 526–29. New York: Oxford University Press.

Black, R. E., L. H. Allen, Z. A. Bhutta, L. E. Caulfield et al. 2008. Maternal and child undernutrition: Global and regional exposures and health consequences. *The Lancet* 371: 243–60.

Blackburn, D. G. 1993. Lactation: Historical patterns and potential for manipulation. *Journal of Dairy Science* 46: 3195–212.

Blurton Jones, N. 1993. The lives of hunter-gatherer children: Effects of parental behavior and parental reproductive strategy. In *Juvenile Primates: Life History, Development, and Behavior*, ed. M. E. Pereira and L. A. Fairbanks, 309–26. New York: Oxford University Press.

Blurton Jones, N., K. Hawkes, and J. F. O'Connell. 1999. Some current ideas about the evolution of the human life history. In *Comparative Primate Socioecology*, vol. 22, *Cambridge Studies in Biological Anthropology*, ed. P.C. Lee, 140–66. Cambridge: Cambridge University Press.

Boinski, S. 1988. Sex differences in the foraging behavior of squirrel monkeys in a seasonal habitat. *Behavioral Ecology and Sociobiology* 32: 177–86.

Bronson, F. H. 1985. Mammalian reproduction: An ecological perspective. *Biology of Reproduction* 32: 1–26.

Brown, K. H., N. A. Akhtar, A. D. Robertson, and M. G. Ahmed. 1986. Lactational capacity of marginally nourished mothers: Relationships between maternal nutritional status and quantity and proximate composition of milk. *Pediatrics* 78: 909–19.

Buss, D. H. 1971. Mammary glands and lactation. In *Comparative Reproduction of Nonhuman Primates*, ed. E. S. E. Hafez, 315–33. Springfield, IL: Charles C. Thomas.

Butte, N. F. 1996. Energy requirements of infants. *European Journal of Clinical Nutrition* 50: S24–S36.

Butte, N. F., C. J. K. Henry, and B. Torun. 1996. Report of the working group on energy requirements of infants, children and adolescents. *European Journal of Clinical Nutrition* 50: S188–S189.

Charnov, E. L., and D. Berrigan. 1993. Why do female primates have such long lifespans and so few babies? Or life in the slow lane. *Evolutionary Anthropology* 1: 191–94.

Cifelli, R. L., T. B. Rowe, W. P. Luckett, J. Banta et al. 1996. Fossil evidence for the origin of the marsupial pattern of tooth replacement. *Nature* 379: 715–18.

Conklin-Brittain, N., R. Wrangham, and C. C. Smith. 2002. A two-stage model of increased dietary quality in early Hominid evolution: The role of fibre. In *Human Diet: Its Origin and Evolution. Human Diet: Perspectives on its Origin and Evolution*, ed. P. S. Ungar and M. F. Teaford, 661–676. Westport, CT: Bergin & Garvey.

Coutsoudis, A. 2005. Breastfeeding and HIV. *Best Practice & Research Clinical Obstetrics and Gynaecology* 19: 185–96.

Daan, S., and J. M. Tinbergen. 1997. Adaptation of life histories. In *Behavioral Ecology: An Evolutionary Approach*, 4th ed., ed. J. R. Krebs and N. B. Davies, 311–33. Oxford: Blackwell.

Dall, S. R. X., and I. L. Boyd. 2004. Evolution of mammals: Lactation helps mothers to cope with unreliable food supplies. *Proceedings of the Royal Society of London: Series B Biological Sciences* 271: 2049–57.

Darwin, C. 1872. *On the Origin of Species*. London: John Murray.

Dettwyler, K. A. 2004. When to wean: Biological versus cultural perspectives. *Clinical Obstetrics and Gynecology* 47: 712–23.

Dewey, K. G., R. J. Cohen, K. H. Brown, and L. L. Rivera. 2001. Effects of exclusive breastfeeding for four versus six months on maternal nutritional status and infant motor development: Results of two randomized trials in Honduras. *Journal of Nutrition* 131: 262–67.

Dewey, K. G., C. A. Lovelady, L. A. Nommsen, M. A. McCrory et al. 1994. A randomized study of the effects of aerobic exercise by lactating women on breastmilk volume and composition. *New England Journal of Medicine* 330: 449–53.

Dufour, D. L., and M. L. Sauther. 2002. Comparative and evolutionary dimensions of the energetics of human pregnancy and lactation. *American Journal of Human Biology* 14: 584–602.

Dunbar, R. I. M., and P. Dunbar. 1988. Maternal time budgets of gelada baboons. *Animal Behavior* 36: 970–80.

Dunbar, R. I. M., L. Hannah-Stewart, and P. Dunbar. 2002. Forage quality and the costs of lactation for female gelada baboons. *Animal Behaviour* 64: 801–5.

Durnin, J. V., F. M. McKillop, S. Grant, and G. Fitzgerald. 1985. Is nutritional status endangered by virtually no extra intake during pregnancy? *Lancet* 2: 823–25.

Durnin, J. V. G. A. 1987. Energy requirements of pregnancy: An integration of the longitudinal data from the five-country study. *Lancet* 2: 1131.

Edmond, K. M., C. Zandoh, M. A. Quigley, S. Amenga-Etego et al. 2006. Delayed breastfeeding initiation increases risk of neonatal mortality. *Pediatrics* 117: e380–86.

Fewtrell, M. S. 2003. The long-term benefits of having been breast-fed. *Current Paediatrics* 14: 97–103.

Foley, R. 1995. Evolution and adaptive significance of hominid behaviour. In *Motherhood in Human and Nonhuman Primates*, ed. C. R. Pryce, R. D. Martin, and D. Skuse. Basel: Karger, 27–36.

Fomon, S. J. 1986. Breast-feeding and evolution. *Journal of the American Dietary Association* 86: 317–18.

Food and Nutrition Board 1989. Energy. *Recommended Dietary Allowances*. 10th ed. National Research Council. Washington, DC: National Academy Press, 24–38.

Ford, C. S. 1945. *A Comparative Study of Human Reproduction*, vol. 32, *Yale University Publications in Anthropology.* New Haven, CT: Yale University Press.

Fouts, H. N., B. S. Hewlett, and M. E. Lamb. 2005. Parent-offspring weaning conflicts among the Bofi farmers and foragers of central Africa. *Current Anthropology* 46: 29–50.

Fouts, H. N., and M. E. Lamb. 2005. Weanling emotional patterns among the Bofi foragers of Central Africa: The role of maternal availability. In *Hunter-gatherer Childhoods: Evolutionary, Developmental, and Cultural Perspectives,* ed. B. S. Hewlett and M. E. Lamb, 309–21. Piscataway, NJ: Aldine Transaction.

Futuyma, D. J. 1998. The evolution of life histories. In *Evolutionary Biology,* 3rd ed., ed. D. J. Futuyma, 561–78. Sunderland, MA: Sinauer Associates, Inc.

Galdikas, B. M. F., and J. W. Wood. 1990. Birth spacing patterns in humans and apes. *American Journal of Physical Anthropology* 83: 185–91.

Garza, C. 2006. New growth standards for the 21st century: A prescriptive approach. *Nutrition Reviews* 64: S72–S91.

Gebo, D.L. 2004. A shrew-sized origin for primates. *American Journal of Physical Anthropology* 47: 40–62.

Goldman, A. S. 2001. Breastfeeding lessons from the past century. *Pediatric Clinics of North America* 48: xxiii–xxv.

———. 2007. The immune system in human milk and the developing infant. *Breastfeeding Medicine: The Official Journal of the Academy of Breastfeeding Medicine* 2: 195–204.

Gonzalez-Cossio, T., J-P. Habicht, K. M. Rasmussen, and H. L. Delgado. 1998. Impact of food supplementation during lactation on infant breast-milk intake and on the proportion of infants exclusively breast-fed. *Journal of Nutrition* 128: 1692–1702.

Gunnlaugsson, G., and J. Einarsdottir. 1993. Colostrum and ideas about bad milk—A case-study from Guinea-Bissau. *Social Science and Medicine* 36: 283–88.

Hammel, E. A. 1996. Demographic constraints on population growth of early humans. *Human Nature* 7: 217–55.

Hartmann, P., S. Morgan, and P. Arthur. 1986. Milk letdown and the concentration of fat in breast milk. In *Human Lactation 2: Maternal and Environmental Factors,* ed. M. Hamosh and A. S. Goldman, 275–81. New York: Plenum Press.

Hartmann, P. E., and P. G. Arthur. 1986. Assessment of lactation performance in women. In *Human Lactation 2: Maternal and Environmental Factors,* ed. M. Hamosh and A. S. Goldman, 215–30. New York: Plenum Press.

Hartmann, P. E., S. Rattigan, C. G., L. Prosser Saint et al. 1984. Human lactation: Back to nature. In *Physiological Strategies in Lactation,* vol. 51, *Symposia of the Zoological Society of London,* ed. M. Peaker, R. G. Vernon, and C. H. Knight, 337–68. London: Academic Press.

Harvey, P. H., and T. H. Clutton-Brock. 1985. Life history variation in primates. *Evolution* 39: 559–81.

Harvey, P. H., R. D. Martin, and T. H. Clutton-Brock. 1987. Life histories in comparative perspective. In *Primate Societies,* ed. B. B. Smuts, D. L. Cheney, R. M. Seyfarth, R. Wrangham, and T. T. Struhsaker, 181–96. Chicago: University of Chicago Press.

Hawkes, K. 2006a. Life history theory and human evolution: A chronicle of ideas and findings. In *The Evolution of Human Life History,* ed. K. Hawkes and R. L. Paine, 45–94. Santa Fe, NM: School of American Research Press.

———. 2006b. Slow life histories and human evolution. In *The Evolution of Human Life History,* ed. K. Hawkes and R. L. Paine, 95–126. Santa Fe, NM: School of American Research Press.

Hawkes, K., and R. L. Paine, eds. 2006. *The Evolution of Human Life History. School of American Research Advanced Seminar Series.* Santa Fe, NM: School of American Research.

Hayssen, V. D., and D. G. Blackburn. 1985. Alpha-lactalbumin and the evolution of lactation. *Evolution* 39: 1147–49.

Heinig, M. J. 2001. Host defense benefits of breastfeeding for the infant: Effect of breastfeeding duration and exclusivity. *Pediatric Clinics of North America* 48: 105–23.

Heinig, M. J., L. A. Nommsen, J. M. Peerson, B. Lonnerdal et al. 1993. Intake and growth of breast-fed and formula-fed infants in relation to the timing of introduction of complementary foods: The DARLING study. *Acta Paediatrica* 82: 999–1006.

Herring, D. A., S. R. Saunders, and M. A. Katzenberg. 1998. Investigating the weaning process in past populations. *American Journal of Physical Anthropology* 105: 425–39.

Hewlett, B. S. 1991. Demography and childcare in preindustrial societies. *Journal of Anthropological Research* 47: 1–37.

———. 2005. Introduction: Who cares for hunter-gatherer children? In *Hunter-gatherer Childhoods: Evolutionary, Developmental, and Cultural Perspectives,* ed. B. S. Hewlett and M. E. Lamb, 175–76. Piscataway, NJ: Aldine Transaction.

Hill, K., and A. M. Hurtado. 1995. *Ache Life History: The Ecology and Demography of a Foraging People. Evolutionary Foundations of Human Behavior: An Aldine de Gruyter Series of Texts and Monographs.* New York: Aldine de Gruyter.

———. 1996. *Ache Life History: The Ecology and Demography of a Foraging People.* New York: Aldine Transaction.

Hill, K., and H. Kaplan. 1999. Life history traits in humans: Theory and empirical studies. *Annual Review of Anthropology* 28: 397–430.

Hinde, K. 2006. Milk composition varies in relation to the presence and abundance of Balantidium coli in the mother in captive rhesus macaques (*Macaca mulatta*). *American Journal of Primatology* 69: 625–34.

———. 2007. Milk composition varies in relation to the presence and abundance of Balantidium coli in the mother in captive rhesus macaques (*Macaca mulatta*). *American Journal of Primatology* 69: 1–10.

Horn, H. S. 1978. Optimal tactics of reproduction and life history. In *Behavioral Ecology: An Evolutionary Approach,* 1st ed., ed. J. R. Krebs and N. B. Davies, 272–94. Oxford: Blackwell.

Huffman, S. L., and L. H. Martin. 1994. First feedings: Optimal feeding of infants and toddlers. *Nutrition Research* 14: 127–59.

Institute of Medicine (US) Subcommittee on Nutrition during Lactation. 1991a. Milk composition. In *Nutrition during Lactation*, 113–52. Washington, DC: National Academy of Sciences.

———. 1991b. Milk volume. In *Nutrition during Lactation*, 80–112. Washington, DC: National Academy of Sciences.

Jones, G., R. W. Steketee, R. E. Black, Z. A. Bhutta et al. 2003. The Bellagio Child Survival Study Group (2003). How many child deaths can we prevent this year? *The Lancet* 362: 65–71.

Kanazawa, A. T., T. Miyazawa, H. Hirono, M. Hayashi et al. 1991. Possible essentiality of docosahexaenoic acid in Japanese monkey neonates: Occurrence in colostrum and low biosynthetic capacity in neonate brains. *Lipids* 26: 53–57.

Kaplan, H., K. Hill, A. M. Hurtado, and J. Lancaster. 2001. The embodied capital theory of human evolution. In *Reproductive Ecology and Human Evolution*, ed. P. T. Ellison, 293–317. New York: Aldine de Gruyter.

Kaplan, H., K. Hill, J. Lancaster, and A. M. Hurtado. 2000. A theory of human life history evolution: Diet, intelligence, and longevity. *Evolutionary Anthropology* 9: 156–84.

Kennedy, G. E. 2005. From the ape's dilemma to the weanling's dilemma: Early weaning and its evolutionary context. *Journal of Human Evolution* 48: 123–45.

Knott, C. D. 2001. Female reproductive ecology of the apes: Implications for human evolution. In *Reproductive Ecology and Human Evolution*, ed. P. T. Ellison, 429–63. New York: Aldine de Gruyter.

Konner, M. 2005. Hunter-gatherer infancy and childhood: The !Kung and others. In *Hunter-gatherer Childhoods: Evolutionary, Developmental, and Cultural Perspectives*, ed. B. S. Hewlett and M. E. Lamb, 19–64. Piscataway, NJ: Aldine Transaction.

Konner, M. J. 1977. Infancy among the Kalahari Desert San. In *Culture and Infancy*, ed. P. H. Leiderman, S. R. Tulkin, and A. Rosenfeld, 69–109. New York: Academic Press.

Kramer, M., and R. Kakuma 2002. The optimal duration of exclusive breastfeeding: A systematic review. In *Cochrane Database of Systematic Reviews*, vol. 1. Geneva: World Health Organization.

Kramer, M. S., T. Guo, R. W. Platt, Z. Sevkovskaya et al. 2003. Infant growth and health outcomes associated with 3 compared with 6 mo of exclusive breastfeeding. *American Journal of Clinical Nutrition* 78: 291–95.

Kumar, S., and B. Hedges 1998. A molecular timescale for vertebrate evolution. *Nature* 392: 917–20.

Labbok, M., D. Clark, and A. Goldman. 2005. Breastfeeding: Maintaining an irreplaceable immunological resource. *Breastfeeding Review* 13: 15–22.

Labbok, M. H. 2001. Effects of breastfeeding on the mother. *Pediatric Clinics of North America* 48: 143–58.

Lamb, M. E., and B. S. Hewlett. 2005. Reflections on hunter-gatherer childhoods. *Hunter-gatherer Childhoods: Evolutionary, Developmental, and Cultural Perspectives*. B. S. Hewlett and M. E. Lamb, 407–15. Piscataway, NJ: Aldine Transaction.

Lawrence, M., F. Lawrence, W. A. Coward, T. J. Cole et al. 1987. Energy requirements of pregnancy in the Gambia. *Lancet* 2: 1072.

Lawrence, R. A. 1989. *Breastfeeding: A Guide for the Medical Profession.* 3rd ed. St. Louis: Mosby.

Lee, P. C. 1987. Nutrition, fertility and maternal investment in primates. *Journal of Zoology, London* 213: 409–22.

———. 1996. The meanings of weaning: Growth, lactation, and life history. *Evolutionary Anthropology* 5: 87–96.

———. 1999. Comparative ecology of postnatal growth and weaning among haplorhine primates. In *Comparative Primate Socioecology,* vol. 22, *Cambridge Studies in Biological Anthropology,* ed. P. C. Lee, 111–39. Cambridge: Cambridge University Press.

Lee, P. C., and J. E. Bowman. 1995. Influence of ecology and energetics on primate mothers and infants. In *Motherhood in Human and Nonhuman Primates,* ed. C. R. Pryce, R. D. Martin, and D. Skuse, 47–58. Basel: Karger.

Lee, P. C., P. Majluf, and I. J. Gordon. 1991. Growth, weaning and maternal investment from a comparative perspective. *Journal of Zoology* 225: 99–114.

Leigh, S. R., and G. Blomquist. 2007. Life history. In *Primates in Perspective,* ed. C. J. Campbell, A. Fuentes, K. C. MacKinnon, M. Panger, and S. K. Bearder, 396–407. Oxford: Oxford University Press.

Lovelady, C.A., B. Lönnerdal, and K. G. Dewey. 1990. Lactation performance of exercising women. *American Journal of Clinical Nutrition* 52: 103–9.

Luckett, W. P. 1993. An ontogenetic assessment of dental homologies in therian mammals. In *Mammal Phylogeny,* vol. 1, *Mesozoic Differentiation, Multituberculates, Monotremes, Early Therians, and Marsupials.* F. S. Szalay, M. J. Novacek, and M. C. McKenna, 182–204. New York: Springer.

Lunn, P. G. 1985. Maternal nutrition and lactational infertility: The baby in the driving seat. In *Maternal Nutrition and Lactational Infertility,* ed. J. Dobbing, 41–64. Vevey/New York: Nestlé Nutrition/Raven Press.

Luo, Z. X., A. W. Crompton, and A. L. Sun. 2002. A new mammal form from the early Jurassic and evolution of mammalian characteristics. *Science* 292: 1535–40.

Mace, R. 2000. Review: Evolutionary ecology of human life history. *Animal Behavior* 59: 1–10.

Marquis, G. S. and J. Habicht. 2000. Breast feeding and stunting among toddlers in Peru. In *Short and Long Term Effects of Breast Feeding on Child Health,* ed. B. Koletzko, K. F. Michaelsen, and O. Hernell, 163–72. New York: Kluwer Academic.

Martin, R. D. 1984. Scaling effects and adaptive strategies in mammalian lactation. *Symposia of the Zoological Society of London* 51: 87–117.

McCrory, M. A. 2000. Aerobic exercise during lactation: Safe, healthful, and compatible. *Journal of Human Lactation* 16: 95–98.

———. 2001. Does dieting during lactation put infant growth at risk? *Nutrition Reviews* 59: 18–27.

McGuire, E. 2005. *An Exploration of How Mother's Milk Protects the Infant.* East Malvern, VIC: Lactation Resource Centre, Australian Breastfeeding Association.

Messer, M., A. S. Weiss, D. C. Shaw, and M. Westerman. 1998. Evolution of the monotremes: Phylogenetic relationship to marsupials and eutherians, and estimation of divergence dates based on a-lactalbumin amino acid sequences. *Journal of Mammalian Evolution* 5: 95–105.

Milligan, L. A., and R. P. Bazinet. 2008. Evolutionary modifications of human milk composition: evidence from long-chain polyunsaturated fatty acid composition of anthropoid milks. *Journal of Human Evolution* 55 (6): 1086–95.

National Academy of Sciences. 1989. *Nutrition and Diarrheal Diseases Control in Developing Countries.* Washington, DC: National Academy Press.

Nelson, E. A. S., L. M. Yu, S. Williams, and the. I. C. C. P. S. G. Members. 2005. International child care practices study: Breastfeeding and pacifier use. *Journal of Human Lactation* 21: 289–95.

Nerlove, S. B. 1974. Women's workload and infant feeding practices: A relationship with demographic implications. *Ethnology* 13: 207–14.

Nicolosi, R. J., and R. D. Hunt. 1979. Dietary allowances for nutrients in nonhuman primates. In *Primates in Nutritional Research,* ed. K. C. Hayes, 11–37. New York: Academic Press.

Nishida, T., N. Corp, M. Hamai, T. Hasegawa et al. 2003. Demography, female life history, and reproductive profiles among the chimpanzees of Mahale. *American Journal of Primatology* 59: 99–121.

Nishida, T., H. Ohigashi, and K. Koshimizu. 2000. Tastes of chimpanzee plant foods. *Current Anthropology* 41: 431–38.

Nishida, T., and L. A. Turner. 1996. Food transfer between mother and infant chimpanzees of the Mahale Mountains National Park, Tanzania. *International Journal of Primatology* 17: 947–68.

Oftedal, O.T. 1984a. Body size and reproductive strategy as correlates of milk energy output in lactating mammals. *Acta Zoologica Fennica* 171: 183–86.

———. 1984b. Milk composition, milk yield and energy output at peak lactation: A comparative review. *Symposia of the Zoological Society of London* 51: 33–85.

———. 1986. Milk intake in relation to body size. In *The Breastfed Infant: A Model for Performance,* ed. R. Laboratories, 44–47. Columbus, OH: Ross Laboratories.

———. 2002a. The mammary gland and its origin during synapsid evolution. *Journal of Mammary Gland Biology & Neoplasia* 7: 225–52.

———. 2002b. The origin of lactation as a water source for parchment-shelled eggs. *Journal of Mammary Gland Biology & Neoplasia* 7: 253–66.

Oftedal, O. T., and S. J. Iverson. 1987. Hydrogen isotope methodology for measurement of milk intake and energetics of growth in suckling young. In *Marine Mammal Energetics. Society for Marine Mammalogy Special Publications,* ed. A. C. Huntley, D. P. Costa, G. A. J. Worthy and M. A. Castellini, 67–96. Lawrence, KS: Allen Press.

Paine, R. R., and K. Hawkes. 2006. Introduction. In *The Evolution of Human Life History,* ed. R. L. Paine and K. Hawkes, 3–16. Santa Fe, NM: School of American Research Press.

Patino, E. M., and J. T. Borda 1997. The composition of primate's milks and its importance in selecting formulas for hand rearing. *Laboratory Primate Newsletter* 36: 8–9.

Peaker, M. 2002. The mammary gland in mammalian evolution: A brief commentary on some of the concepts. *Journal of Mammary Gland Biology & Neoplasia* 7: 347–53.

Pennington, R. L. 1996. Causes of early human population growth. *American Journal of Physical Anthropology* 99: 259–74.

Pereira, M. E., and L. A. Fairbanks, eds. 2002. *Juvenile Primates: Life History, Development, and Behavior.* Chicago: University of Chicago Press.

Pérez-Escamilla, R., S. Segura-Millán, J. Canahuati, and H. Allen. 1996. Prelacteal feedings are negatively associated with breast-feeding outcomes in Honduras. *Journal of Nutrition* 126: 2765–73.

Plaut, K., and K. L. Carraway. 2002. Evolution and comparative biology of the mammary gland. *Journal of Mammary Gland Biology and Neoplasia* 7: 223–24.

Pond, C. M. 1997. The biological origins of adipose tissue in humans. In *The Evolving Female,* ed. M. E. Morbeck, A. Galloway, and A. L. Zihlman, 47–162. Princeton, NJ: Princeton University Press.

Power, M. L., O. T. Oftedal, and S. D. Tardif. 2002. Does the milk of callitrichid monkeys differ from that of larger anthropoids? *American Journal of Primatology* 56: 117–27.

Prentice, A. 1986. The effect of maternal parity on lactational performance in a rural African community. In *Human Lactation 2: Maternal and Environmental Factors,* ed. M. Hamosh and A. S. Goldman, 165–73. New York: Plenum Press.

Prentice, A. M., A. A. Paul, A. Prentice, A. K. Black et al. 1986. Cross-cultural differences in lactational performance. In *Human Lactation 2: Maternal and Environmental Factors,* ed. M. Hamosh and A. S. Goldman, 13–43. New York: Plenum Press.

Prentice, A., C. Spaaij, G. Goldberg, S. Poppitt et al. 1996. Energy requirements of pregnant and lactating women. *European Journal of Clinical Nutrition* 50: S82–S111.

Prentice, A., R. Whitehead, S. Roberts, and A. Paul. 1981. Long-term energy balance in childbearing Gambian women. *American Journal of Clinical Nutrition* 34: 2790–99.

Quandt, S. 1985. Biological and behavioral predictors of exclusive breastfeeding duration. *Medical Anthropology* 9: 139–51.

Rasmussen, K. M. 1992. The influence of maternal nutrition on lactation. *Annual Review of Nutrition* 12: 103–17.

Rizvi, N. 1993. Issues surrounding the promotion of colostrum feeding in rural Bangladesh. *Ecology of Food and Nutrition* 30: 27–38.

Roberts, S. B., T. J. Cole, and W. A. Coward. 1985. Lactational performance in relation to energy intake in the baboon. *American Journal of Clinical Nutrition* 41: 1270–76.

Robson, S. L. 2004. Breast milk, diet, and large human brains. *Current Anthropology* 45: 419–24.

Robson, S. L., C. P. van Schaik, and K. Hawkes. 2006. The derived features of human life history. In *The Evolution of Human Life History,* ed. K. Hawkes and R. L. Paine, 17–44. Santa Fe, NM: School of American Research Press.

Sangild, P. T. 2006. Gut responses to enteral nutrition in preterm infants and animals. *Experimental Biology and Medicine* 231: 1–16.

Sauther, M. L. 1994. Changes in the use of wild plant foods in free-ranging ring-tailed lemurs during pregnancy and lactation: Some implications for human foraging strategies. In *Eating on the Wild Side: The Pharmacologic, Ecologic and Social Implications of Using Noncultigens*, ed. N. L. Etkin, 240–46. Tucson: University of Arizona Press.

Schwarcz, H. P. and L. E. Wright. 1998. Stable carbon and oxygen isotopes in human tooth enamel: Identifying breastfeeding and weaning in prehistory. *American Journal of Physical Anthropology* 106: 1–18.

Sellen, D. W. 2001a. Comparison of infant feeding patterns reported for nonindustrial populations with current recommendations. *Journal of Nutrition* 131: 2707–15.

———. 2001b. Weaning, complementary feeding, and maternal decision making in a rural east African pastoral population. *Journal of Human Lactation* 17: 233–44.

———. 2002. Sub-optimal breast feeding practices: Ethnographic approaches to building 'baby friendly' communities. *Advances in Experimental Medicine & Biology* 503: 223–32.

———. 2006. Lactation, complementary feeding and human life history. In *The Evolution of Human Life History*, ed. K. Hawkes and R. L. Paine, 155–97. Santa Fe, NM: School of American Research Press.

———. 2007. Evolution of infant and young child feeding: implications for contemporary public health. *Annual Review of Nutrition* 27: 123–48.

———. 2009. Evolution of human lactation and complementary feeding: Implications for understanding contemporary cross-cultural variation. *Breast-feeding: Early Influences on Later Health*. Series: *Advances in Experimental Medicine and Biology*, vol. 369, ed. G. Goldberg, A. Prentice, A. Prentice, S. Filteau et al., 253–82. New York, Springer.

Sellen, D. W., and R. Mace. 1999. A phylogenetic analysis of the relationship between subadult mortality and mode of subsistence. *Journal of Biosocial Science* 31: 1–16.

Sellen, D. W., and D. B. Smay. 2001. Relationship between subsistence and age at weaning in pre-industrial societies. *Human Nature: An Interdisciplinary Journal* 12: 47–87.

Smith, B. H. 1991. Dental development and the evolution of life history in Hominidae. *American Journal of Physical Anthropology* 86: 157–74.

———. 1992. Life history and the evolution of human maturation. *Evolutionary Anthropology* 1: 134–42.

Stearns, S. C. 1992. *The Evolution of Life Histories*. Oxford: Oxford University Press.

Sugarman, M., and K. Kendall-Tackett. 1995. Weaning ages in a sample of American women who practice extended breastfeeding. *Clinical Pediatrics* 34: 642–47.

Tardif, S. D., M. L. Harrison, and M. A. Simek. 1993. Communal infant care in marmosets and tamarins: Relation to energetics, ecology, and social organization. In *Marmosets and Tamarins: Systematics, Behaviour, and Ecology*, ed. A. B. Rylands, 220–34. New York: Oxford University Press.

Tilden, C. D. and O. T. Oftedal. 1997. Milk composition reflects pattern of maternal care in prosimian primates. *American Journal of Primatology* 41: 195–211.

Trevathan, W. R., and J. J. McKenna. 1994. Evolutionary environments of human birth and infancy: Insights to apply to contemporary life. *Children's Environments* 11: 88–104.

Underwood, B. A., and Y. Hofvander. 1982. Appropriate timing for complementary feeding of the breast-fed infant. *Acta Paediatrica Scandinavica* S294: 5–32.

van Esterik, P. 2002. Contemporary trends in infant feeding research. *Annual Review of Anthropology* 31: 257–78.

van Noordwijk, M.A., and C. P. van Schaik. 2005. Development of ecological competence in Sumatran orangutans. *American Journal of Physical Anthropology* 127: 79–94.

van Schaik, C.P., N. Barrickman, M. L. Bastian, E. B. Krakauer et al. 2006. Primate life histories and the role of brains. In *The Evolution of Human Life History,* ed. K. Hawkes and R. L. Paine, 127–54. Santa Fe, NM: School of American Research Press.

Walker, R., K. Hill, O. Burger, and A. M. Hurtado. 2006. Life in the slow lane revisited: Ontogenetic separation between chimpanzees and humans. *American Journal of Physical Anthropology* 129: 577–83.

Winkvist, A., F. Jalil, J-P. Habicht, and K. M. Rasmussen. 1994. Maternal energy depletion is buffered among malnourished women in Punjab, Pakistan. *Journal of Nutrition* 124: 2376–85.

Wood, J. W. 1990. Fertility in anthropological populations. *Annual Reviews of Anthropology* 19: 211–42.

World Health Organization and UNCS Fund. 2003. *Global Strategy for Infant and Young Child Feeding.* Geneva: World Health Organization.

Wray, J. 1991. Breast-feeding: An international and historical review. In *Infant and Child Nutrition Worldwide: Issues and Perspectives,* ed. F. Falkner, 62–117. Boca Raton: CRC Press.

Wright, P. 1990. Patterns of paternal care in primates. *International Journal of Primatology* 11: 89–102.

Breastfeeding and Beyond: Nutrition throughout the Life Course

• 4 •
The Use of Stable Isotope Analysis to Determine Infant and Young Child Feeding Patterns

T. L. Dupras

Introduction

The reconstruction of infant[1] and young child feeding (IYCF) patterns can provide valuable information in the quest for understanding the behaviors of past populations. How a population treats its very young and the cultural ideologies regarding IYCF practices can be a key determinant of infant and child survival and this factor can heavily influence overall population growth or decline (Delgado et al. 1982; Stuart-Macadam and Dettwyler 1995; Lewis 2007). Therefore, the study of IYCF practices in past populations is recognized as a significant component in the reconstruction of population demography (Herring et al. 1998). Although there have been a few skeletal markers postulated as potentially indicating the onset of the weaning process (e.g., linear enamel hypoplasias, cribra orbitalia, Harris lines), the cause of these skeletal indicators may be multiple, and not necessarily solely the result of the weaning process. Because of the unknown etiology of these skeletal markers, other methods may provide more reliable indicators of the infant weaning process.

An accepted methodology used to determine diet and IYCF patterns from the skeletal remains of archaeological populations is the analysis of stable isotopes. Studies of this kind have been conducted on archaeological samples from various time periods and regions around the world including populations from Africa (e.g., White and Schwarcz 1994; Dupras 1999; Dupras et al. 2001; White et al. 2004; Clayton et al. 2006; Turner et al. 2007), Europe (e.g., Dittmann and Grupe 2000; Richards et al. 2002; Prowse et al. 2005; Fuller et al. 2005; Bourbou and Richards 2007), North America (e.g., Katzenberg et al. 1993; 1995; Schurr 1997; Schurr and Powell 2005), Central America (e.g., Wright and Schwarcz 1998; 1999), and South America (e.g., Williams et al. 2005). These studies have revealed the timing and pattern of the transitional feeding period and weaning in different societies and the possible repercussions these patterns could have had within these societies.

Stable Isotopes and Infant Feeding Patterns

At present, investigations of IYCF patterns in archaeological samples generally rely on the use of three stable isotopes: carbon ($\delta^{13}C$), nitrogen ($\delta^{15}N$), and oxygen ($\delta^{18}O$). Traditionally IYCF studies focus on isotopic signals from bone collagen (as opposed to bone hydroxyapatite), as it is possible to look at both carbon and nitrogen from this component of the bone. The collagen portion of the bone reflects the protein portion of the diet, while bone mineral (hydroxyapatite) reflects the total diet. As this chapter focuses on the uses of stable isotope analysis for determining infant and young child diet, please refer to DeNiro (1987), Katzenberg (1992, 2000), and Schwarcz and Schoeninger (1991) for extended reviews of the uses of stable isotope analyses in reconstructing general adult diet in archaeological contexts.

Carbon Isotopes

Carbon stable isotopes are commonly used to distinguish between diets based on C_3 and C_4 plants[2] (e.g., Katzenberg et al. 1995). C_3 plants, such as wheat, barley, rice, grasses, trees, and most fruits and vegetables, range in $\delta^{13}C$ values from -22‰ to -33‰,[3] with a mean of -28‰ (Smith and Epstein 1971). C_4 plants are enriched in ^{13}C in comparison to C_3 plants with values that range from -16‰ to -9‰, with a mean of -13.5‰ (Smith and Epstein 1971), and include plants such as maize, sorghum, some millets, sugar cane, and tropical grasses. The carbon isotope signals of plants are transferred through the food web, and so there is a direct relationship between the isotopic values of the plant and the consumer. There is approximately a +5‰ difference between plant $\delta^{13}C$ values and those of human collagen (Ambrose and Norr 1993) due to physiological fractionation of isotopes in the human body. Thus, humans consuming a diet based on C_3 plants will have collagen $\delta^{13}C$ values of approximately -19‰. Humans consuming a diet consisting only of C_4 plants will have collagen $\delta^{13}C$ values of approximately -8‰. Individuals that consume a mixed diet of both C_3 and C_4 plants have both collagen and bone/enamel carbonate $\delta^{13}C$ values that are intermediate between the above values.

The transfer of carbon isotopes through the food chain is important to understand when interpreting diet. Because breastfeeding infants are at the top of the food chain, we also have to consider that there are most likely physiological fractionation processes that impact isotope ratios in their bodies. Recent literature (Richards et al. 2002; Richards et al. 2006) has demonstrated that breastfeeding infants in archaeological contexts have $\delta^{13}C$ values that are enriched by approximately 1‰ over adult female values. This chemical signal has been called the "carnivore" effect of breastfeeding, in which carnivore collagen is enriched in comparison to that of herbivore collagen—in this case the infant being the carnivore as it consumes its mother's milk (Schoeninger and DeNiro 1984; Bocherens et al. 1995). Fuller and colleagues (2006) have shown that modern infants who are exclusively breastfeeding have a 1‰ increase in $\delta^{13}C$ values, adding further support to the carnivore effect. This signal can be useful when interpreting the timing of the weaning process, as it indicates that the infant is most likely only consuming breast milk and not complementary foods when the 1‰ increase in $\delta^{13}C$ values is apparent. However, as will be demonstrated later, caution must be taken when interpreting the weaning process in some archaeological populations. It may be possible that complementary foods that include C_4 plants (e.g., cereal made from millet or sorghum, or milk consumed from animals that are fed C_4 plants) may also cause enrichment in the $\delta^{13}C$ value (Dupras et al. 2001; Dupras and Tocheri 2007), thus potentially causing a misinterpretation of the timing of the introduction of complementary foods.

Nitrogen Isotopes

Nitrogen stable isotopes are used to determine the timing of the weaning process by illustrating the shift between a reliance on human milk proteins to those obtained from alternative foods (e.g., Fogel et al. 1989; Katzenberg and Pfeiffer 1995; Schurr 1998). Reconstructions of IYCF patterns commonly rely on nitrogen isotope analyses from bone collagen (Katzenberg and Pfeiffer 1995; Schurr 1998; Schurr and Powell 2005), and the isotopic composition of nitrogen in collagen reflects the isotopic composition of dietary proteins. There is a stepwise increase, known as the trophic level effect, of approximately 3‰ from one level of the food chain to the next (Schoeninger and DeNiro 1984), and because a breastfeeding infant receives its protein from its natural (or surrogate) mother, the $\delta^{15}N$ value of the infant's tissues is approximately 2–3‰ higher than that of its mother (Fogel et al. 1989). Breastfeeding infants will have the highest $\delta^{15}N$ values in the population because they are at the top of the food chain. The typical weaning process is gradual as complementary foods are added slowly to the infant's diet as dependency on breast milk decreases; this dietary change is detected through decreasing $\delta^{15}N$ values that eventually reach adult values once the infant is weaned and infants are completely dependent on exogenous food sources (Schurr 1998; Dupras 1999; Dupras et al. 2001; Dupras and Tocheri 2007). Studies utilizing nitrogen isotopes have revealed variations in the timing of the weaning process depending on the population studied. For example, abrupt weaning at the age of 1.5 is reported for individuals from a nineteenth-century Methodist cemetery in Newmarket, Ontario (Katzenberg and Pfeiffer 1995), while most others report a long transitional feeding period with foods being slowly introduced. Weaning ages of between two and five years have been reported in a sample from prehistoric Ohio (Schurr and Powell 2005), and as old as six years in Middle Preclassic through Late Postclassic (ca. 700 BCE to 1500 CE) Guatemala (Wright and Schwarcz 1998). Richards et al. (2002) and Fuller and colleagues (2003) report weaning at the medieval English site of Wharram Percy to have occurred around two years of age, while White and Schwarcz (1994) report a weaning age of three to four years for ancient Nubian (Meroitic to Christian period, ca. 350 BCE to 1400 CE) subadults. From a biocultural perspective, the timing of weaning may have been an important determinant of the demographics of the population, with cultural ramifications.

Oxygen Isotopes

Oxygen stable isotope analyses are used to show patterns of water intake. For example, stable oxygen isotopes of dental enamel from various modern and fossil animals have shown seasonal changes in water consumption (Koch et al. 1992; Stuart-Williams and Schwarcz 1997; Balasse 2002). Studies of oxygen isotopes in archaeological contexts are commonly used to detect patterns of migration as shifts in geographical locations will cause a change in available water sources, and water sources are marked by differences in oxygen values resulting from various factors (e.g., altitude, temperature, etc.). These differences are transferred to the consumer and are reflected in differing oxygen isotope signals from different tissue types. Oxygen stable isotopes are also used in the determination of IYCF patterns to detect the shift in water consumption from that of breast milk to environmental sources (Wright and Schwarcz 1998). The weaning process is detectable through shifts in $\delta^{18}O$ because breast milk is significantly enriched in ^{18}O compared with drinking water. Thus, as an infant survives the weaning process, their $\delta^{18}O$ values change due to the inclusion of other water sources in their diet. Once the infant or child is weaned, the values begin to reach similar levels as adults in the population. This "weaning signal" has been documented both in extant and extinct animals (Fricke and O'Neil 1996; Franz-Odendaal et al. 2003), as well as in archaeological human populations (Wright and Schwarcz 1998; White et al. 2004; Williams et al. 2005). The stable oxygen isotope signal of enamel apatite carbonate closely reflects the $\delta^{18}O$ values of water in the body. Enamel is a static tissue; therefore, the isotopic signature captured during dental crown development remains constant throughout an individual's life. Since each tooth in the dental arcade develops at a different time during growth (Hillson 2003), it is possible to examine the $\delta^{18}O$ values of teeth that develop during infancy to detect when exogenous water sources are introduced and when the weaning process ends (Wright and Schwarcz 1998; Dupras and Tocheri 2007).

The Dakhleh Oasis, Egypt: A Case Study

The Dakhleh Oasis in Egypt, located approximately 450 miles southwest of Cairo (Figure 4.1), has been under archaeological investigation by the Dakhleh Oasis Project (DOP) since 1978 (Mills 1984). One of the main foci of the DOP has been to understand how humans came to live and survive in such a harsh environment, with extreme variations in temperature, little precipitation, and other environmental hardships such as sand storms. One of the ways in which we can understand human survival and adaptation in this environment is to study the human remains of those buried in the oasis. The ancient village of Kellis (see inset of Figure 4.1 for location within the Dakhleh Oasis), occupied during the Roman-Christian period[4], has been under archaeological investigation since 1984. Although there are tombs located within the village of Kellis, the main burial sites appear to be the two cemeteries that flank the village to the west and east. Even though archaeological excavations and skeletal analysis have been conducted on the tombs within the village (Dupras and Tocheri 2004a, 2004b), and on the Kellis 1 (west) cemetery (Birrell 1999), the Kellis 2 (east) cemetery has received the most analytical attention.

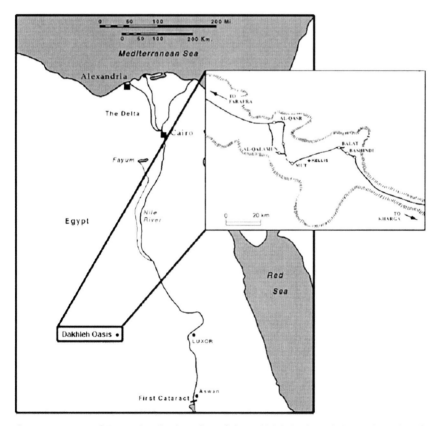

Figure 4.1. Map of Egypt showing location of the Dakhleh Oasis and the ancient site of Kellis (see inset).

Excavations and mortuary analysis of the Kellis 2 cemetery have revealed a Christian-style burial practice (Birrell 1999; Bowen 2003). With few exceptions, each individual (including fetuses and infants), has their own burial. Individuals are placed in an east–west orientation with the head facing east. Skeletons are typically found in an extended position, with the hands lying either beside or on top of the pelvis. Although extensive artificial mummification was no longer practiced on those interred in this cemetery, there is evidence that many individuals in this cemetery received some mortuary treatment such as clay mixed with botanicals, oils, and myrrh (Williams and Dupras 2004). The arid environment and alkaline soil in the Dakhleh Oasis also helped to naturally preserve textiles and the soft tissue of many of these individuals (Figure 4.2, inset). Although grave inclusions are rare, mainly due to ancient and modern looting, some individuals were buried with items such as botanicals, jewelry, and pottery.

At present, 654 individuals have been excavated from the Kellis 2 cemetery. Bioarchaeological investigations of the Kellis 2 cemetery includes ongoing research in areas such as mortuary treatment (Birrell 1999; Williams and Dupras 2004), paleodemography and migration (Dupras and Schwarcz 2001), ancient diet and nutrition (Dupras 1999; Dupras et al. 2001; Dupras and Tocheri 2007), skeletal growth and development (Tocheri and Molto 2002; Tocheri et al. 2005), changes in health and disease (Fairgrieve and Molto 2000; Maggiano et al. 2003; Maggiano et al. 2006; Molto 2000, 2001, 2002), and paleogenetics (Graver et al. 2001; Parr 2002; Stewart et al. 2003). The skeletal sample excavated from this cemetery is very important for the analysis of infant and child remains, as 450 of the 654 skeletons that have been excavated are juveniles (see Figure 4.2). Of these, 114 are classified as fetal/perinate, 143 as infants, and 153 as children (up to the age of fifteen years) (see Table 4.1). Thus, the sample is large enough to make significant statements about infant weaning and feeding practices in Roman period Dakhleh Oasis.

Infant and Young Child Diet in the Dakhleh Oasis

Dupras (1999) and Dupras and colleagues (2001) conducted initial stable carbon and nitrogen isotope analyses on a cross-sectional sample of infants and children from the Kellis 2 cemetery. The purpose of this study was to detect the feeding and weaning pattern of this population and to determine if this pattern was similar to that documented in historical Roman literature. This study included the analysis of bone collagen from forty-nine individuals that ranged in age from birth to eight years. In addition, Dupras (1999) and Dupras et al. (2001) also analyzed carbon and nitrogen from archaeological botanical and nonhuman animal remains in order to reconstruct the isotopic food-web.

Initial isotopic results indicated that the infants from the Kellis 2 cemetery were breastfed exclusively for the first six months of life, with a slow transitional feeding period, and weaning by three years of age. The carbon and nitrogen isotope results from this sample showed complementary results. Figure 4.3A shows the combination

Figure 4.2. Excavation map from the Kellis 2 cemetery in the Dakhleh Oasis, Egypt. Inset shows infant burials from the Kellis 2 cemetery, illustrating the preservation of burial wrappings and soft tissue.

Table 4.1. Number of individuals in each age category from the Kellis 2 cemetery. Abbreviations for age categories include: (w) = fetal weeks; (m) = months after 40 fetal weeks; and (y) = years.

Age Category	Age Range	Sample size
Fetal	0–36 w	24
Perinate	37–40 w	90
Infant	0–12 m	143
Young Child	1–4 y	89
Middle Child	5–10 y	49
Older Child	11–15 y	17
Adult	16–19 y	12
Adult	20–35 y	112
Adult	36–50 y	72
Adult	51–59 y	25
Adult	60 y+	21
Total		654

of mean $\delta^{13}C$ and $\delta^{15}N$ isotope data from this study (Dupras et al. 2001), with the addition of data from the subsequent analysis of thirty-five more individuals in the same age range. The pattern of nitrogen data (appearing on the left side of the graph, with data plotted as the solid line) shows that infant and child $\delta^{15}N$ values increased to a peak value at six months of age, reaching a trophic level of approximately 3‰ over adult females (20.8‰[5] for infants at six months of age versus a mean of 18‰ for adult females). This indicates an exclusive dependence on breast milk until approximately six months of age. Nitrogen values then begin a steady decrease until approximately three years of age, when childhood values approach those of adult females.

The $\delta^{13}C$ values (Figure 4.3A) also show an increase during the first year of life, reaching values that are approximately 1‰ enriched over those of adult females. This enrichment, however, occurs at a slightly older age than the peak values of nitrogen. Although current research suggests a carnivore effect that occurs in breastfeeding infants (Fuller et al. 2006; Richards et al. 2006), Dupras and colleagues (2001) suggest that the enrichment in $\delta^{13}C$ values in this site maybe due to the types of complementary foods that were being introduced at approximately six months of age. The archaeological presence of pearl millet (*Pennisetum glaucum*) in Kellis (Thanheiser 1999), and isotopic evidence from cow and goat remains, suggests that a C_4 signal may have been present in the diet of these infants (Thanheiser et al. 2007). Isotopic

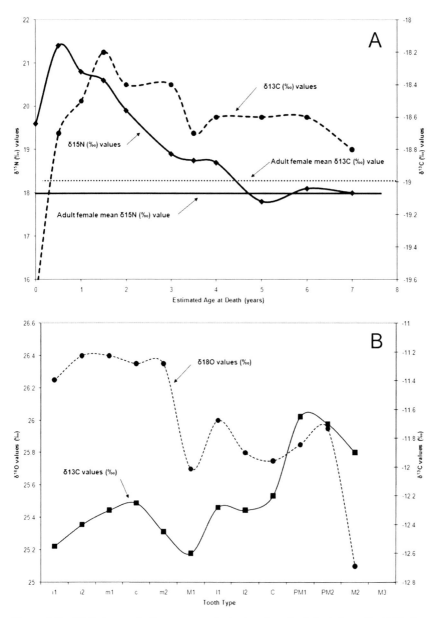

Figure 4.3. (A) Carbon and nitrogen mean isotopic values for infants from the Kellis 2 cemetery. The solid data line with squares represents the mean nitrogen values, while the dashed line with circles represents the carbon values. Corresponding mean values for the adult females are indicated by the straight lines with the solid line representing nitrogen (mean $\delta^{15}N$ is 18‰) and the dashed line representing carbon (mean $\delta^{13}C$ is −19‰). (B) Enamel oxygen and carbon mean isotopic values of teeth from individuals from the Kellis 2 cemetery. The oxygen isotopic data are represented by a dashed line, while the carbon data are represented by a solid line.

analysis of nonhuman animal remains from Kellis revealed that both cows and goats were significantly enriched in ^{13}C in comparison to other animals, suggesting that these animals were fed mostly millet as fodder. Ancient literary sources such as Galen and Soranus suggested that mothers should begin introducing complementary foods at six months of age, and that gruel/pap mixtures made with cow or goat milk was favored (Tempkin 1951; Green 1956). Given this evidence, it is completely within the realm of possibility that the infants of Kellis were fed complementary foods that either consisted of pearl millet or that they drank milk from cows that were fed pearl millet, thereby indirectly receiving the C_4 signal. The combination of complementary foods, and the possibility of physiological fractionation, i.e., the carnivore effect, gives us clues as to when the weaning process began in this population, after six months of age and slowly returning to adult values by three years.

Cross-sectional versus Longitudinal Isotopic Studies of Infant Diet at the Dakhleh Oasis

Is the story of infant and young child diet in ancient Kellis finished? Not yet. Dupras and Tocheri (2007) considered another aspect of IYCF that is commonly overlooked in most studies. Almost all isotopic studies of IYCF practices, even that of Dupras et al. (2001) described previously, are designed in such a manner that the data represent a cross-sectional IYCF profile. It is important to recognize that there may be a mortality bias (Wood et al. 1992) that has a direct impact on the interpretation of IYCF practices in past populations. To state the obvious, infants and young children studied from cemetery samples had the misfortune of dying early, and the possibility that these individuals were sick, and most likely not thriving, is high. From a sociocultural perspective, it is important to consider that these infants and young children may have been fed different diets than their healthy, thriving counterparts (Stuart-Macadam and Dettwyler 1995), or may have even had differences in their isotopic fractionation due to illness (Williams 2007a; 2007b). To be able to determine if IYCF patterns detected through cross-sectional studies are in fact representative of that society's practices, designing a study that simulates longitudinal data is important. One way in which to do this is to analyze a sample of individuals who survived the weaning process (e.g., those older than three years), and still have tissues present that were formed during the weaning process.

As mentioned previously, teeth are the perfect tissue to use for the longitudinal reconstruction of diet because they are static structures that once formed do not change throughout life (Hillson 2003) and therefore capture isotopic signals, giving a snapshot of diet during that period. Since the formation of human dentition begins during gestation and does not cease until approximately twenty years of age, it is possible to detect patterns of diet and weaning from teeth. Dupras and Tocheri (2007) analyzed a total of 297 teeth from 102 juvenile and adult individuals from the Kellis 2 cemetery in order to determine if the cross-sectional weaning data reported by Dupras et al. (2001) showed the same pattern as that reconstructed from the longitudinal profile. Dupras and Tocheri (2007) analyzed $\delta^{13}C$ and $\delta^{18}O$ from the dental enamel and $\delta^{15}N$ and $\delta^{13}C$ from the dentin to detect changes indicative of IYCF practices.

When comparing deciduous and permanent dentition, Dupras and Tocheri (2007) found significant differences between the $\delta^{13}C$ enamel values of the permanent dentition and those of the deciduous teeth, with the permanent dental enamel being statistically significantly enriched by 0.3‰ over that of the deciduous teeth (adult dentition = -12.1‰ versus deciduous dentition = -12.4‰) (Figure 4.4A). Earlier-forming permanent dentition (I1, I2, C, and M1) were also found to be significantly enriched by 0.6‰ over later-forming permanent teeth (P1, P2, and M2). As illustrated in Figure 4.4A, dental enamel forms first, and the timing of the formation of enamel varies from tooth to tooth. Therefore when we consider the timing of dental formation in conjunction with the isotopic data, the interpretation of the isotopic data becomes clear. The formation of the deciduous dentition commences during fetal development, with the first and second incisor, and first molar crowns almost completely forming before birth. Thus, the enamel $\delta^{13}C$ values of the deciduous teeth mostly reflect the fetal environment and are most likely a direct reflection of maternal $\delta^{13}C$ values. An examination of Figure 4.3B shows that there is very little isotopic variation between the deciduous and early-forming permanent teeth. Dupras and Tocheri (2007) note that the difference in enrichment in the enamel $\delta^{13}C$ values derives mainly from the later-forming permanent teeth (P1, P2, and M2). Because the crowns of all the permanent teeth begin forming after birth, and continue forming until approximately seven years of age (Figure 4.4A), the $\delta^{13}C$ enrichment of the permanent teeth includes ^{13}C signals from breastfeeding, transitional feeding, and complete adoption of adult foods. However, a closer inspection of Figure 4.3B shows that ^{13}C enrichment actually begins with the formation of the permanent dentition, but mostly occurs during the formation of the later-forming permanent teeth, all of which are primarily forming after two years of age. This, as indicated earlier in the study by Dupras et al. (2001), could signify the introduction and increased reliance on foods that are C_4 enriched, such as pearl millet, or indirectly through cow and/or goat milk, and these foods may be responsible for the ^{13}C enrichment shown in those teeth forming from birth to approximately three years of age.

When examining enamel $\delta^{18}O$ values Dupras and Tocheri (2007) found that the deciduous dentition was significantly enriched (0.5‰) over that of the permanent dentition (26.3‰ versus 25.8‰, respectively) (Figure 4.3B). Because changes in water source consumption are reflected by changes in $\delta^{18}O$ values in the consumer's tissues (Wright and Schwarcz 1998), we can see the process of weaning reflected through shifts in $\delta^{18}O$ because breast milk is significantly enriched in ^{18}O in comparison to drinking water, so as an infant changes from only breast milk to other sources, its ^{18}O values will change. The isotopic signatures of the deciduous tooth crowns reflect the $\delta^{18}O$ values captured during fetal growth and the first year of life, and thus reflect the uterine environment and breast milk. Because both of these water sources

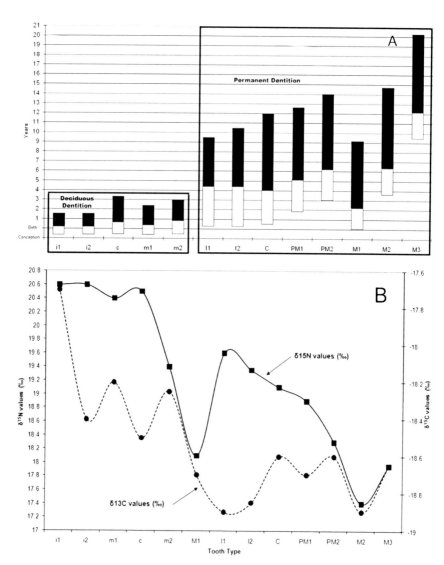

Figure 4.4. (A) Age of formation for the mandibular deciduous and permanent dentition. Each tooth type is listed on the X-axis in order of appearance in the mouth from medial to distal. Crown formation is indicated by the white bar, while root formation is indicated in black. Data compiled from Moorrees *et al.* (1963) and Smith (1991). (B) Dentin nitrogen and carbon isotopic values of teeth from individuals from the Kellis 2 cemetery. The nitrogen data are represented by a solid line, while the carbon data are represented by a dashed line.

come directly from a maternal source, we would expect that enamel isotopic signatures should be enriched in $\delta^{18}O$, as seen in the Kellis 2 sample. Permanent dental enamel forms after birth (Figure 4.4A), with almost 50 percent of crown development after two years of age. Thus, the $\delta^{18}O$ values of these teeth (Figure 4.3B) reflect both breastfeeding and the adoption of other water sources, with the majority of the signal coming from external water sources.

Dupras and Tocheri (2007) also note significant enrichment (0.6‰) of deciduous dentin $\delta^{13}C$ values over permanent dentition (-18.2‰ versus -18.8‰, respectively) (Figure 4.4B). Deciduous dentin develops after birth during breastfeeding, transitional feeding, and weaning, later than the formation of dental enamel, which forms mostly prior to birth. The deciduous roots and the crowns of the early-forming permanent teeth form around the same time, and thus show a similar enrichment level. Dentin begins to form in most of the permanent teeth (with the exception of M1) after the third year of development; therefore, the ^{13}C enrichment of the deciduous dentin is likely due to the breastfeeding and weaning process. By the time the M1 root starts to develop (after three years of age), the mean ^{13}C values have already decreased to adult values. Although this enrichment could be due to the carnivore effect (Fuller et al. 2006; Richards et al. 2006), we cannot exclude the possibility that the enrichment is in fact due to the inclusion of complementary foods that are enriched in C_4. Dupras and Tocheri (2007) argue that the evidence from Kellis (e.g., textual evidence as mentioned previously, existence of C_4 plants, and cow and goat $\delta^{13}C$ values) should be considered when interpreting $\delta^{13}C$ values in this population.

Deciduous dentin $\delta^{15}N$ values of the individuals from the Kellis 2 cemetery are significantly enriched (2‰) over that of the permanent dentition (20‰ versus 18‰, respectively) (Dupras and Tocheri 2007). This is caused by the 2–3‰ trophic level effect that occurs when infants are breastfeeding and only receive their protein from a maternal source (Fogel et al. 1989). Because the dentin of the deciduous teeth forms after birth and is mainly complete by three years of age (Figure 4.4A), the enriched isotopic signal indicates that infants from Kellis were breastfeeding during this time, but the process had ended by three years of age. The comparison of early- versus late-forming permanent dentition shows that early-forming permanent dentition (I1, I2, C, and M1) were significantly enriched by 0.5‰, suggesting a gradual weaning process.

Although it is not possible to get at specific ages for the introduction of complementary foods because the dental isotopic data are based on homogenized tooth samples that span long periods of time, the longitudinal dental isotope data suggests that individuals were introduced to a food enriched in ^{13}C during the first two years of life, and that the weaning process was complete by three years of age. Both the longitudinal study by Dupras and Tocheri (2007) and the cross-sectional study by Dupras and colleagues (2001) provide complementary evidence as to the nature and timing of IYCF at Kellis. Together, the results of the isotopic analyses from Kellis indicate that infants and young children were breastfed exclusively for six months and then were in the transitional feeding phase until three years of age, which is also consistent with traditional IYCF practices documented in ancient texts by Soranus and Galen (Green 1951; Tempkin 1956). Other historical sources from Egypt also suggest a transitional feeding period of three years (Bagnall et al. 2005; Fildes 1986; Robins 1993; Donadoni 1997). The comparison of two data sources from the Kellis 2 cemetery suggests that an infant mortality bias may not be a significant problem for stable isotope analyses in this population.

Conclusion

The interpretation of IYCF patterns can make a significant contribution to the reconstruction of past populations. It can give us clues about child-rearing practices, including information such as the interaction between childhood nutrition, pathology, and environmental change (Lewis 2007). Although IYCF practices have the potential to have an impact on the overall demography of a population, hence potentially contributing to the overall well-being or decline of that population, the data from this population suggest that many individuals successfully survived through the weaning process. There does not appear to have been differential treatment of sick versus thriving infants in terms of IYCF practices.

Since the end of the 1980s stable isotope analysis, particularly carbon, nitrogen, and oxygen isotopes, has been used as a reliable technique in the reconstruction of IYCF practices in many archaeological populations around the world. The particular archaeological population highlighted in this chapter, that of the Kellis 2 cemetery in the Dakhleh Oasis, Egypt, demonstrates the utility of stable isotopes in the reconstruction of IYCF practices. The interpretation of IYCF in the Kellis 2 population will also be combined with other biocultural/bioarchaeological information from the same population to gain a greater understanding of IYCF practices in Roman Egypt.

Acknowledgements

I would like to acknowledge the Supreme Council of Antiquities in Egypt for their continued support of the DOP. In addition I would like to thank particular members of the DOP for their support and contributions to this research—T. Mills, P. Sheldrick, M. Tocheri, and E. Molto. Special thanks to S. Wheeler and L. Williams for sharing the results of their research—it has helped immensely in the interpretation of the juvenile remains from this site.

Notes

1. The use of the term *infant and young child feeding* (IYCF) follows Sellen (chapter 3, this volume). Also following Sellen, *transitional feeding* is defined as "the phase during which

nutrition is derived from a combination of maternal milk and other foods foraged by the infant, its parents, or others", and *weaning* is the termination of suckling. See Sellen (chapter 3, Table 3.1) for these definitions and others. Unlike Sellen, I use the term *weaning process* to denote the beginning of the transitional feeding stage that ends with the termination of suckling, or weaning.
2. C_3 and C_4 are used to differentiate the physiological pathways in which plants metabolize carbon from the environment. C_3 plants use the enzyme bisphosphate decarboxylase to fix atmospheric carbon, resulting in a compound with three carbon atoms, while C_4 plants utilize the enzyme phosphoenol pyruvate carboxylase, resulting in a compound with four carbon atoms, thus C_4 plants are more enriched in carbon.
3. Stable isotope values are expressed using the δ notation where $δ = [(R_{sample}/R_{standard})-1] \times 1000$. $R = {}^{13}C/{}^{12}C$ for $δ^{13}C$; $R = {}^{15}N/{}^{14}N$ for $δ^{15}N$; $R = {}^{18}O/{}^{16}O$ for $δ^{18}O$. All δ values are expressed in parts per mille (thousand) (‰).
4. The archaeological context and C^{14} dates for this site are not congruent, and are still under investigation. At this time the archaeological evidence suggests that the village of Kellis was occupied from approximately 50 to 390 CE (Hope 2001; Bowen 2003), while C^{14} dates from the Kellis 2 cemetery show that the cemetery may have been in use from as early as 40 CE to as late as 540 CE (Molto pers. comm.; Molto 2001; Stewart et al. 2003).
5. It should be noted that the nitrogen values from this site are substantially elevated (similar to those documented for marine-based diets) in comparison to data from other populations known to have subsisted on a terrestrial diet. This phenomenon (also noted by Aufderheide et al. 1988) occurs in desert environments where the base of the food chain has enriched nitrogen values (see Schwarcz and colleagues 1999 for further explanation).

References

Ambrose, S. H., and L. Norr. 1993. Experimental evidence for the relationship of carbon isotope ratios of whole diet and dietary protein to those of bone collagen and carbonate. In *Prehistoric Human Bone Archaeology at the Molecular Level*, ed. J. B. Lambert and G. Grupe, 1–37. Berlin, Springer-Verlag.

Aufderheide, A. C., L. L. Tieszen, M. J. Allison, J. Wallgren et al. 1988. Chemical reconstruction of components in complex diets: A pilot study. In *Diet and Subsistence: Current Archaeological Perspectives*, ed. B. V. Kennedy and G. M. LeMoine, 301–6. Calgary: University of Calgary Archaeological Association.

Bagnall, R. S., C. Helms, and A. Verhoogt. 2005. Documents from Berenike II: Texts from the 1999–2001 seasons. Brussels, Association Egyptologique Reine Elisabeth. *Papyrologica Bruxelensia* 33: 41–43.

Baker, B. J., T. L. Dupras, and M. W. Tocheri. 2005. *The Osteology of Infants and Children*. College Station: Texas A&M Press.

Balasse, M. 2002. Reconstructing dietary and environmental history from enamel isotopic analysis: Time resolution of intra-tooth sequential sampling. *International Journal of Osteoarchaeology* 12 (3): 155–65.

Birrell, M. 1999. Excavations in the cemeteries of Ismant el-Kharab. In *Dakhleh Oasis Project: Preliminary Reports on the 1992–1993 and 1993–1994 Field Seasons*, ed. C.A. Hope and A.J. Mills, 29–41. Oxford: Oxbow Books.

Bocherens, H., M. L. Fogel, N. Tuross, and M. Zender. 1995. Trophic structure and climatic information from isotopic signatures in Pleistocene cave fauna of southern England. *Journal of Archaeological Science* 22: 327–40.

Bourbou, C., and M. P. Richards 2007. The Middle Byzantine menu: Palaeodietary information from isotopic analyses of humans and fauna from Kastella, Crete. *International Journal of Osteoarchaeology* 17: 63–72.

Bowen, G. E. 2003. Some observations on Christian burial practices at Kellis. In *The Oasis Papers 3: The Proceedings of the Third International Conference of the Dakhleh Oasis Project*, ed. G. E. Bowen and C. A. Hope, 166–82. Oxford: Oxbow Books.

Clayton, F., J. Sealy, and S. Pfeiffer. 2006. Weaning age among foragers at Matjes River Rock Shelter, South Africa, from stable nitrogen and carbon isotope analyses. *American Journal of Physical Anthropology* 129: 311–17.

Delgado, H. L., R. Martorell, and R. E. Klein. 1982. Nutrition, lactation, and birth interval components in rural Guatemala. *American Journal of Clinical Nutrition* 35: 1468–76.

DeNiro, M. J. 1987. Stable isotopy and archaeology. *American Scientist* 75: 182–91.

Dittmann, K., and G. Grupe. 2000. Biochemical and palaeopathological investigations on weaning and infant mortality in the early Middle Ages. *Anthropologischer Anzeiger* 58: 345–55.

Donadoni, S. 1997. *The Egyptians*. Chicago: University of Chicago Press.

Dupras, T. L. 1999. "Dining in the Dakhleh Oasis, Egypt: Determination of Diet using Documents and Stable Isotope Analysis." PhD dissertation, McMaster University.

Dupras, T. L., and H. P. Schwarcz. 2001. Strangers in a strange land: Stable isotope evidence for human migration in the Dakhleh Oasis, Egypt. *Journal of Archaeological Science* 28: 1199–208.

Dupras, T. L., H. P. Schwarcz, and S. I. Fairgrieve. 2001. Infant feeding and weaning practices in Roman Egypt. *American Journal of Physical Anthropology* 115: 204–11.

Dupras, T. L., and M. W. Tocheri. 2004a. A preliminary analysis of the human skeletal remains from North Tomb 1 and 2. In *The Oasis Papers 3: The Proceedings of the Third International Conference of the Dakhleh Oasis Project*, ed. G. E. Bowen and C. A. Hope, 183–96. London: Oxbow Press.

———. 2004b. Preliminary Analyses of Human Skeletal Remains from South Tomb 4. In *The Oasis Papers 3: The Proceedings of the Third International Conference of the Dakhleh Oasis Project*, ed. G. E. Bowen and C. A. Hope, 197–99. London: Oxbow Press.

———. 2007. Reconstructing infant weaning histories at Roman period Kellis, Egypt using stable isotope analysis of dentition. *American Journal of Physical Anthropology* 134: 63–74.

Fairgrieve, S. I., and J. E. Molto. 2000. Cribra orbitalia in two temporally disjunct population samples from the Dakhleh Oasis, Egypt. *American Journal of Physical Anthropology* 111: 319–30.

Fildes, V. 1986. *Breast, Bottles and Babies: A History of Infant Feeding.* Edinburgh: Edinburgh University Press.
Fogel, M. L., N. Tuross, and D. Owsley. 1989. Nitrogen isotope tracers of human lactation in modern and archaeological populations. In *Annual reports of the Director, Geophysical Laboratory, 1988–1989,* 111–16. Washington, DC: Carnegie Institute of Washington.
Franz-Odendaal, T. A., J. A. Lee-Thorp, and A. Chinsamy. 2003. Insights from stable isotopes on enamel defects and weaning in Pliocene herbivores. *Journal of Bioscience* 28(6): 765–73.
Fricke, H. C., and J. R. O'Neil. 1996. Inter- and intra-tooth variations in the oxygen isotope ratio of mammalian tooth enamel phosphate: Implications for paleoclimatological and paleobiological research. *Palaeogeography, Palaeoclimatology, Palaeoecology* 126: 91–100.
Fuller, B. T., T. I. Molleson, D. A. Harris, L. T. Gilmour et al. 2005. Isotopic evidence for breastfeeding and possible adult dietary differences from late/Sub-Roman Britain. *American Journal of Physical Anthropology* 129: 45–54.
Fuller, B. T., J. L. Fuller, D. A. Harris, and R. E. M. Hedges. 2006. Detection of breastfeeding and weaning in modern human infants with carbon and nitrogen stable isotope ratios. *American Journal of Physical Anthropology* 129: 279–93.
Graver, A. M., J. E. Molto, R. L. Parr, S. Walters et al. 2001. Mitochondrial DNA research in the Dakhleh Oasis, Egypt: A preliminary report. *Ancient Biomolecules* 3: 239–53.
Green, R. M., trans. 1951. *Galen. Hygiene (De Sanitate tuenda).* Springfield, IL: Thomas.
Herring, D. A., S. R. Saunders, and M. A. Katzenberg. 1998. Investigating the weaning process in past populations. *American Journal of Physical Anthropology* 105 (4): 425–40.
Hillson, S. 2003. *Dental Anthropology.* Cambridge: Cambridge University Press.
Hope, C. A. 2001. Observations on the Dating of the Occupation at Ismant el-Kharab. In *The Oasis Papers I: The Proceedings of the First Conference of the Dakhleh Oasis Project,* ed. C. A. Marlow and A. J. Mills, 43–59. Oxford: Oxbow Books.
Katzenberg, M. A. 1992. Advances in stable isotope analysis of prehistoric bones. In *Skeletal Biology of Past Peoples: Research Methods,* ed. S. R. Saunders and M. A. Katzenberg, 105–19. New York: John Wiley & Sons, Inc.
———. 2000. Stable Isotope Analysis: A Tool for Studying Past Diet, Demography and Life History. In *Biological Anthropology of the Human Skeleton,* ed. M. A. Katzenberg and S. R. Saunders, 305–27. New York: John Wiley & Sons, Inc.
Katzenberg, M. A., and S. Pfeiffer. 1995. Nitrogen isotope evidence for weaning. In *Bodies of Evidence: Reconstructing History through Skeletal Analysis,* ed. A. L. Grauer, 221–35. New York: John Wiley & Sons, Inc.
Katzenberg, M. A., S. R. Saunders, and W. Fitzgerald. 1993. Age differences in stable carbon and nitrogen isotope ratios in a population of prehistoric maize horticulturists. *American Journal of Physical Anthropology* 90: 267–81.
Katzenberg, M. A., H. P. Schwarcz, M. Knyf, and F. J. Melbye. 1995. Stable isotope evidence for maize horticulture and paleodiet in southern Ontario, Canada. *American Antiquity* 60: 335–50.

Koch P. L., J. C. Zachos, and P. D. Gingerich. 1992. Correlation between isotope records in marine and continental carbon reservoirs near the Palaeocene/Eocene boundary. *Nature* 358: 319–22.

Lewis, M. E. 2007. *The Bioarchaeology of Children: Perspectives from Biological and Forensic Anthropology.* Cambridge: Cambridge University Press.

Maggiano, C., T. L. Dupras, and J. Biggerstaff. 2003. Ancient antibiotics: Evidence of tetracycline in human and animal bone from Kellis. In *The Oasis Papers 3: The Proceedings of the Third International Conference of the Dakhleh Oasis Project*, ed. G. E. Bowen and C. A. Hope, 331–44. Oxford: Oxbow Books.

Maggiano, C., T. L. Dupras, M. Schultz, and J. Biggerstaff. 2006. Spectral and photobleaching analysis using confocal laser scanning microscopy: A comparison of modern and archaeological bone fluorescence. *Molecular and Cellular Probes* 20: 154–62.

Mills, A. J. 1984. Research in the Dakhleh Oasis. In *Origin and Early Development of Food Processing Cultures in North-Eastern Africa*, ed. L. Kryzaniak and M. Kobusiewicz, 205–10. Poznan: Polish Academy of Sciences.

Molto, J. E. 2000. Humerus varus deformity in Roman period burials from Kellis 2, Dakhleh, Egypt. *American Journal of Physical Anthropology* 113: 103–9.

———. 2001. The comparative skeletal biology and paleoepidemiology of the people from 'Ein Tirghi and Kellis, Dakhleh Oasis, Egypt. In *The Oasis Papers 1: The Proceedings of the First International Symposium of the Dakhleh Oasis Project*, ed. M. Marlow and A. J Mills, 81–100. Oxford: Oxbow Books.

———. 2002. Bio-archaeological research of Kellis 2: An overview. In *Dakhleh Oasis Project: Preliminary Reports on the 1994–1995 to 1998–1999 Field Seasons*, ed. C. A. Hope and G. E. Bowen, 239–55. Oxford: Oxbow Books.

Moorrees, C. F., E. A. Fanning, and E. E. Hunt. 1963. Formation and resorption of three deciduous teeth in children. *American Journal of Physical Anthropology* 21: 205–13.

Parr, R. L. 2002. Mitochondrial DNA sequence analysis of skeletal remains from the Kellis 2 cemetery. In *Dakhleh Oasis Project: Preliminary Reports on the 1994–1995 to 1998–1999 Field Seasons*, ed. C. A. Hope and G. E Bowen, 257–61. Oxford: Oxbow Books.

Prowse, T., H. Schwarcz, S. Saunders, L. Bondioli et al. 2004. Isotopic paleodiet studies of skeletons from the Imperial Roman cemetery of Isola Sacra, Rome, Italy. *Journal of Archaeological Science* 31: 259–72.

Prowse, T., H. Schwarcz, S. Saunders, L. Bondioli et al. 2005. Isotopic evidence for age-related variation in diet from Isola Sacra, Italy. *American Journal of Physical Anthropology* 128: 2–13.

Richards, M. P., B. T. Fuller, and T. I. Molleson. 2006. Stable isotope palaeodietary study of humans and fauna from the multi-period (Iron Age, Viking and Late Medieval) site of Newark Bay, Orkney. *Journal of Archaeological Science* 33: 122–31.

Richards, M. P., S. Mays, and B. T. Fuller. 2002. Stable carbon and nitrogen values of bone and teeth reflect weaning age at the Medieval Wharram Percy Site, Yorkshire, UK. *American Journal of Physical Anthropology* 119: 205–10.

Robins, G. 1993. *Women in Ancient Egypt.* Cambridge MA: Harvard University Press.

Schoeninger, M. J., and M. J. DeNiro. 1984. Nitrogen and carbon isotopic composition of bone collagen from marine and terrestrial animals. *Geochimica e Cosmochimica Acta* 48: 625–39.

Schurr, M. R. 1997. Stable isotopes as evidence for weaning at the Angel Site: A comparison of isotopic and demographic measures of weaning age. *Journal of Archaeological Science* 24: 919–27.

———. 1998. Using stable nitrogen isotopes to study weaning behavior in past populations. *World Archaeology* 30: 327–42.

Schurr, M. R., and M. L. Powell. 2005. The role of changing childhood diets in the prehistoric evolution of food production: An isotopic assessment. *American Journal of Physical Anthropology* 126: 278–94.

Schwarcz, H. P., T. L. Dupras, and S. I. Fairgrieve. 1999. ^{15}N enrichment in the Sahara: In search of a global relationship. *Journal of Archaeological Science* 26: 629–36.

Schwarcz, H. P., and M. J. Schoeninger. 1991. Stable isotope analyses in human nutritional ecology. *Yearbook of Physical Anthropology* 34: 283–322.

Smith, B. H. 1991. Standards of human tooth formation and dental age assessment. In *Advances in Dental Anthropology*, ed. M. A. Kelley and C. S. Larsen, 143–68. New York: Wiley-Liss.

Smith, B. N., and S. Epstein. 1971. Two categories of $^{13}C/^{12}C$ ratios for higher plants. *Plant Physiology* 47: 380–84.

Stewart, J. D., J. E. Molto, and P. Reimer. 2003. The chronology of Kellis 2: The interpretive significance of radiocarbon dating of human remains. In *The Oasis Papers 3: The Proceedings of the Third International Conference of the Dakhleh Oasis Project*, ed. G. E. Bowen and C. A. Hope, 345–64. Oxford: Oxbow Books.

Stuart-Macadam, P., and K. Dettwyler. 1995. *Breastfeeding: Biocultural Perspectives*. New York: Aldine de Gruyter.

Stuart-Williams, H. L. Q., and H. P. Schwarcz. 1997. Oxygen isotopic determination of climatic variation using phosphate from beaver bone, tooth enamel, and dentine. *Geochimica e Cosmochimica Acta* 61: 2539–50.

Tempkin, O., trans. 1956. *Soranus of Ephesus, Gynecology*. Baltimore: Johns Hopkins Press.

Thanheiser, U. 1999. Plant remains from Ismant el-Kharab: first results. In *Dakhleh Oasis Project: Preliminary Reports on the 1992–1993 and 1993–1994 Field Seasons*, ed. C. A. Hope and A. J. Mills, 89–93. Oxford: Oxbow Books.

Thanheiser, U., S. Kahlheber, and T. L. Dupras. in press. The importance of pearl millet (*Pennisetum glaucum*) in the Dakhleh Oasis. In *Oasis Papers IV*, ed. O. Kaper and F. Liemhaus. London: Oxbow Press.

Tocheri, M. W., and J. E. Molto. 2002. Ageing fetal and juvenile skeletons from Roman period Egypt using basiocciput osteometrics. *International Journal of Osteoarchaeology* 12: 356–63.

Tocheri, M. W., T. L. Dupras, P. Sheldrick, and J. E. Molto. 2005. Roman period fetal skeletons from the East Cemetery (Kellis 2) of Kellis, Egypt. *International Journal of Osteoarchaeology* 15: 326–41.

Turner, B. L., J. L. Edwards, E. A. Quinn, J. D. Kingston et al. 2007. Age-related variation in isotopic indicators of diet at Medieval Kulubnarti, Sudanese Nubia. *International Journal of Osteoarchaeology* 17: 1–25.

White, C., F. J. Longstaffe, and K. Law. 2004. Exploring the effects of environment, physiology and diet on oxygen isotope rations in ancient Nubian bones and teeth. *Journal of Archaeological Science* 31: 233–50.

White, C., and H. P. Schwarcz. 1994. Temporal trends in stable isotopes for Nubian mummy tissues. *American Journal of Physical Anthropology* 93: 165–87.

Williams, J. S., C. D. White, and F. J. Longstaffe. 2005. Trophic level and macronutrient shift effects associated with the weaning process in the Postclassic Maya. *American Journal of Physical Anthropology* 128: 781–90.

Williams, L. 2007a. The eleven percent solution? Tissue rehydration as a means of reducing growth-cycle error in isotopic analysis of hair segments. (abstract) *VI Meeting of the World Mummy Congress*. Tenerife, Canary Islands.

———. 2007b. "Investigating seasonality of death at Kellis 2 cemetery using solar orientation and isotopic analysis of mummified tissues." PhD dissertation. London, Ontario, University of Western Ontario.

Williams, L., and T. L. Dupras. 2004. Mortuary mixing: Evidence of body treatment in a Roman/Early Christian cemetery. (abstract) *32nd Annual Meeting of the Canadian Association for Physical Anthropology*. London, Ontario.

Wright, L. E., and H. P. Schwarcz. 1998. Stable carbon and oxygen isotopes in human tooth enamel: identifying breastfeeding in prehistory. *American Journal of Physical Anthropology* 106: 1–18.

———. 1999. Correspondence between stable carbon, oxygen and nitrogen isotopes in human tooth enamel and dentine: Infant diets at Kaminaljuya. *Journal of Archaeological Science* 26: 1159–70.

Wood, J. W., G. R. Milner, H. C. Harpending, and K. M. Weiss. 1992. The osteological paradox. *Current Anthropology* 33: 343–70.

• 5 •
A Community in Transition
Deconstructing Breastfeeding Trends in Gibraltar, 1955–1996

L. A. Sawchuk, E. K. Bryce, and S. D. A. Burke

Introduction

Despite widespread scholarly interest in infant feeding practices, there are few long-term quantitative studies on the subject at the national or community-based level (see, for example, Martinez and Nalezienski 1981; Kinter 1985; McNally, Henericks, and Horowitz 1985; Liestøl, Rosenberg, and Walløe 1988; Siskind, Del Mar, and Schofield 1993; Ryan 1997). Reasons for the paucity of studies are complex, but some factors include: (1) lack of availability of medical and survey-based data on infant feeding, (2) the issue of deconstructing variation existing within and between complex social units at the local and national level, and (3) most importantly, a focus on more immediate outcome measures, such as low infant survivorship and poor infant/child growth and development.

The present study contributes to our understanding of secular change in infant feeding by examining four decades when rapid sociopolitical change was driven by both global and local events. Supported by a quantitatively based cohort construct and supplemented by qualitative information, we posit that rapid change in culturally mediated behavior such as breastfeeding is "best" examined and understood when the importance of time and place and its impact on collective social behavior is recognized.

Situating Infant Feeding Practices: Decades of Previous Research

There exists a staggering quantity of literature on the benefits of breastfeeding including its role in reducing infant mortality and morbidity and improving maternal health. Culturally mediated infant feeding decisions can have immediate and long-term biological consequences. It has been suggested that breastfeeding mediates infant exposure to pathogens, promotes healthy physiological and cognitive development, and protects against adult obesity, heart disease, diabetes, and allergies (Dettwyler and Fishman 1992; Cunningham 1995; Holman and Grimes 2003; Kramer and Kakuma 2002; Horta et al. 2007; but see Bauchner, Leventhal, and Shapiro 1986). Studies in the developing world tend to focus on protection against acute infectious diseases (for example Kramer and Kakuma 2002), while those in developed nations concentrate on long-term benefits (Horta et al. 2007). Because of its reported advantages, breastfeeding is promoted by national governments, advocacy groups, professional medical organizations, and global organizations including the WHO and UNICEF (Gengler, Mulvey, and Oglethorpe 1999).

The specialization of research on infant feeding—from disciplines as diverse as clinical nursing, medicine, public health, nutrition, psychology, sociology, and history—has led to what some authors term the "compartmentalization" of knowledge (Quandt 1998). Since anthropology is located at the intersection of biology and culture, it is well suited to the topic of infant feeding. Anthropologists are able to contribute to the literature by providing a "deep" analysis of a behavior that is both cultural and biological (Van Esterik 2002). By promoting interdisciplinary study (Dettwyler and Fishman 1992), the field moves away from a quest for simple solutions (Quandt 1995).

Supplementing breast milk with other foods has always been a feature of infant feeding (Fildes 1986), but became much more widespread when commercially prepared infant foods appeared by the mid-nineteenth century. In its earliest stages, supplementary feeding was a "death warrant" for babies (Apple 1987: 4). Infants were typically fed with pap, a mixture of bread and water, or cow's milk, both of which are nutritionally insufficient for a human newborn (Apple 1987; Fildes 1986). Faced with high rates of infant mortality due to malnutrition, both the medical profession and mothers needed an acceptable substitute for breast milk. Artificial substitutes were manufactured as early as 1840; by the 1890s many different commercially prepared infant formulas were available throughout Europe and North America (Apple 1994).

The early 1900s was a time of rapid technological innovation. North America became a new frontier for the development of "scientific" infant formula (Apple 1987). At the turn of the twentieth century, physicians were experiencing an increase in prestige and authority. They were the guardians of science, holding knowledge the public did not possess (Shorter 1985). Science represented progress and the creation of complex infant formulas became the "modern" way to feed a baby. As infant formula became more popular, expertise on feeding increasingly became the domain of physicians (Apple 1987). Advances in bacteriology, nutrition, and physiology improved the quality of formula, and artificial feeding became more widespread. The commercial production of infant formula was a highly profitable endeavor, and it was heavily marketed to both women and doctors. By the 1930s, following pressure from doctors, marketing was limited to medical professionals (Apple 1994). Over time, artificial feeding became a socially and medically acceptable substitute for the "old-fashioned" practice of breastfeeding (Apple 1994; Hausman 2003) and, by World War II, breastfeeding rates in many developed countries had plummeted.

From the end of World War II to the early 1970s, when breastfeeding rates reached their lowest point, breastfeeding initiation and duration rates continued to drop steadily in Europe, North America, and Australia (Jelliffe and Jelliffe 1978; Martinez and Nalezienski 1981; Hendershot 1984; Liestøl, Rosenberg, and Walløe 1988; Apple 1994; Ryan 1997). The medicalization of motherhood, which began with artificial formula, continued as birthing moved from the home to the hospital (Apple 1987). Hospitalized births required strict rules for mothers. Due to fear of diarrheal disease, babies were isolated in antiseptic wards except for routine feeding times. Early initiation of breastfeeding and mother-infant bonding was not possible under such a system, and the scheduled nature of breastfeeding likely led to problems of insufficient milk supply (Apple 1994). It is not surprising, then, that mothers would lose confidence in their ability to nourish infants solely by breastfeeding, and turn to artificial feeding as a substitute. The medical profession was also ambivalent about breastfeeding: it was not commonly taught in medical school and, at the time, artificial feeding was believed to be a nutritionally acceptable substitute for breastfeeding (Hirschman and Butler 1981).

The Second World War also marked a change in women's status in society. Women who worked in factories during the war were reluctant to return to the home, and female participation in the labor force grew steadily throughout the next few decades. Disposable incomes rose, and being able to afford "modern" infant formula was a status symbol (Quandt 1995). Women with more education and higher incomes were leaders in changes in infant feeding trends (Liestøl, Rosenberg, and Walløe 1988). Educated women were often more aware of current literature, which at the time promoted artificial feeding at least as actively as breastfeeding (Apple 1987; Van Esterik 1989; Knaak 2005).

After reaching a low in the early 1970s (Ryan, Wenjun, and Acosta 2002; see also Jelliffe and Jelliffe 1978), breastfeeding rates began to reverse their downward trend. The social activism of the 1960s and 1970s had spawned feminist movements that raged against medicalization of infant feeding and advocated a return to natural childbirth and motherhood (Liestøl, Rosenberg, and Walløe 1988; Quandt 1995; Hausman 2003).

By the late 1980s the upward trend in breastfeeding was well established, increasing at a rate of about 2 percent per year in the United States (Ryan, Wenjun, and Acosta 2002). Yet, breastfeeding initiation and duration have not yet met guidelines established by government agencies or the WHO. Many authors, particularly those using an epidemiological approach to identifying risk factors, end their articles with the suggestion that more education will solve the problem. The implicit assumption is that if women just had the proper knowledge, they would automatically make the decision to breastfeed (Ryan 1997; Gengler, Mulvey, and Oglethorpe 1999; Van Esterik 2002). As in earlier time periods (Fildes 1986; Apple 1987; Dobbing 1988; Van Esterik 2002), there is currently little dispute that breastfeeding is the optimal method of infant feeding. Although today almost all women try to initiate breastfeeding at birth (Scott et al. 2001; Ryan, Wenjun, and Acosta 2002; Hofvander 2005; Lanting, van Wouwe, and Reijneveld 2005; Horta et al. 2007; Singh, Kogan, and Dee 2007), many stop soon after leaving the hospital (Liestøl, Rosenberg, and Walløe 1988; Scott et al. 2001). The high rates of breastfeeding initiation suggest that women want to try breastfeeding, but for some reason choose not to continue. In contrast to the beginning of the twentieth century, women who choose to artificially feed their infants today do so not because they are actively embracing bottle feeding, but because they are rejecting breastfeeding (Losch et al. 1995).

The Case Study Approach

The preceding discussion describes infant feeding trends primarily in Britain, Scandinavian countries, the United States, and Australia. The trends are often implied to be universal, although authors do acknowledge regional variation (Ryan 1997). An examination of secular trends in breastfeeding in Gibraltar complements these earlier studies by adding to our knowledge regarding the scope of variation in breastfeeding trends as well as how quickly change can occur in infant feeding practices, a fundamental culturally defined element of society.

The grounding for this research lies in case study methodology. While there are discipline-based variations inherent to the case study approach (see for example Tellis 1997), there are a number of common attributes collectively defining this type of investigation. Most importantly, the fundamental unit of study is the community. In Gibraltar, the community possesses a constellation of special qualities that render it akin to a "living laboratory" for research studies. It is self-defined, spatially bounded, and contains sufficient diversity to allow a changing range of responses over time, including those triggered by sociopolitical forces. Case studies adopt inherently triangulated research strategies—using multiple and independent sources of information encourages the desirable qualities of validity, reliability, and consistency. The case study approach aims for both a holistic perspective, to situate findings in a broader context, and a fine-grained perspective, to understand the peculiarities of local dynamics.

The Study Site: Gibraltar

Gibraltar is a small British overseas territory located at the western entrance of the Mediterranean. Connected to Spain at its northern extremity, Gibraltar is otherwise surrounded by water. The landscape of "the Rock" is dominated by a 1,396-foot limestone outcrop, with most Gibraltarians living in tightly packed housing clusters around its base. The territory is a diminutive 6.5 square kilometers in size and, with a population size of just under 27,000 in 1991, Gibraltarians continued to live under very crowded and congested conditions over the duration of our twentieth-century study period. Agricultural and large-scale industrial undertakings are not possible on the Rock given its limited size and local resources. As a result, Gibraltarians are highly dependent on imports, including basic necessities such as food and petroleum, from other places such as Spain, England, and Morocco. For much of the twentieth century, the civil population was involved in wage earning connected with commercial activities, the civil service, the military establishment, the dockyard, and construction/development projects. When England drew Gibraltar into the European Economic Community (now the European Union) in 1973, it was with the provision that Gibraltar would be a value-added-tax (VAT)-free territory. It is therefore an attractive locale for tourist spending; an estimated 5.5 million tourists entered Gibraltar by land, sea, or air in 1995 alone (Abstract of Statistics 1995).

Within this small-scale community, face-to-face contact of its residents is a daily occurrence. Living in a "cradle-to grave" context, emigration is relatively rare. In recent years, immigration has become more common as people from Britain, the European Union, and Morocco come to live and work long term. Family, neighbors, and the community serve as primary units of social reference in this highly cohesive population. The Gibraltarian culture reflects a unique fusion of Mediterranean, European, and Northern African influences, coupled with the long-term and shared experience of living in a highly overcrowded military garrison and colonial outpost. Some vestiges of colonialism remain evident, including a relatively rigid class structure. Overall, education is relatively homogeneous, though disparities emerge at the postsecondary level since, until recently, access to colleges and universities typically meant the heavy financial commitment of leaving Gibraltar.

All civilian infant births occur at the only local government hospital (St. Bernard's). During the study period, the hospital was centrally located and within minutes of all Gibraltar's residents (it has since moved). Following the English system, health care is subsidized by the government; as a result, the possibility for differential pre- and postnatal care because of relative wealth is minimal. Midwives and nurses follow women through their pregnancies beginning at about three months after conception (Sawchuk, Burke, and Benady 1997).

A Population in Flux: Social Change Under Siege

Gibraltarians have enjoyed many generations of stability on the Rock, though their lives were shaken on two occasions in the twentieth century. The first disruptive force was World War II. As part of the British Government's wartime mandate, all Gibraltarians, except able-bodied men, were evacuated. Though all those wanting to come back to Gibraltar were eventually repatriated, the process was slow and many families spent years waiting to return (see Finlayson 1996). Our study period begins in the 1950s, just as the majority of Gibraltar's 15,700 repatriated citizens had trickled back to the Rock from places as distant as England, Northern Ireland, Madeira, Jamaica, and Tangier (Government of Gibraltar 2001).

Just as the impact of this exceptional event was receding into memory, another disruptive force set in—the politically motivated border closure between Spain and Gibraltar, which began in 1969 and lasted some fifteen years until the border reopened in 1985. Characterized as the fifteenth siege of Gibraltar (Jackson 1987), the border closure was part of a deeply rooted Spanish irredentist campaign to regain sovereignty over Gibraltar after more than two-and-a-half centuries of British rule; this objective intensified during General Franco's dictatorship of Spain (for a thorough discussion on the twentieth-century politics between Spain and Gibraltar, see Gold 1994). The campaign intensified after World War II, as Gibraltarians established local government representation, a situation that Spanish authorities argued was in violation of the Treaty of Utrecht.

Spanish authorities first focused on depriving Gibraltar of Spanish laborers, in an attempt to strangle Gibraltar's local economy. In the words of one Spanish representative: "Economically speaking, Gibraltar cannot live without Spain. There it lives at the cost of Spain and constitutes a sort of cancer in the economy of our country" (PR no. 166/67 1967). For years, Gibraltar had been acknowledged in Spain as "la piedra gorda" (the fat rock) in light of Gibraltar's more prominent economy that allowed better pay and working conditions than in neighboring southern Spain (see Stewart 1967). By withdrawing Spanish laborers and goods, it was hoped that England would feel the burdens of supplying an isolated territory and come to view Gibraltar as an unwise investment. In 1965 restrictions were first placed on Spanish women working in Gibraltar. On 9 June 1969, there was a complete withdrawal of the 4,666 Spanish men who entered Gibraltar daily to work (constituting a full third of Gibraltar's work force). Two weeks later, on 26 June, Gibraltar was completely cut off when Spain's border gates slammed shut. Although life became difficult for Gibraltarians and it was costly for England to provision the isolated community, local feelings of opposition remained ingrained. With both sides remaining obstinate, the border closure continued for fifteen long years. It ended when the post-Franco Spanish democratic government sought membership in the European Community, since such blatant hostility against a neighboring Community member would not be acceptable (PR 30 January 1981).

Gibraltar became a different community during the border closure. Cut off from Europe's mainland, tourism plummeted and the economy suffered. For the first time in Gibraltar's history the community had to learn to live with long-term isolation, effectively becoming an island community. Solidarity was cemented and a family-like feeling reigned as people united in a common cause to resist Spanish coercion (Burke 1999). Yet the border closure had negative repercussions as well. Since provisions could no longer be acquired from Spain, the cost of living rose dramatically. Despite hardships created by the border closure, population health continued to improve, reflected in declining infant mortality rates and increasing life expectancy (see Figure 5.1).

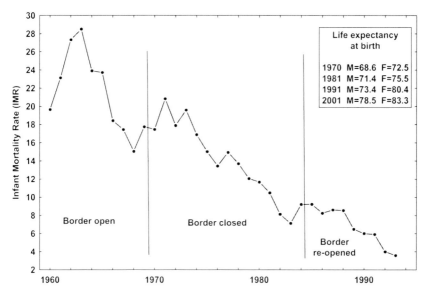

Figure 5.1. Health in Gibraltar, 1960–1993: Infant mortality rate and life expectancy by sex.

During this period, the gendered division of labor changed dramatically. In response to the shortage of workers from nearby Spain, Gibraltarian women entered the workforce in unprecedented numbers. Former restrictions on women's work outlining the kind of work women could be expected to perform were necessarily modified or removed (Burke 1999). Despite the need for female labor power, and despite the equal pay ordinance of 1975, women's jobs remained poorly paid and lacking in prestige (Martens 1987). Women in the community continued to depend on men as primary breadwinners (Martens 1987). When the border reopened in 1985, it resulted in a social and economic transformation in Gibraltar (Sawchuk 1992), as if the residents were being "pulled out of the Victorian age" (Burke 1999).

Materials and Methods

The local hospital's maternity discharge register provided summary information on mothers and infants (see Sawchuk, Burke, and Benady 1997 for a more complete description). Since Gibraltar has only one maternity facility, our data are not a sampling of births, but represent the entire population (though some births were excluded from analysis, as noted below). Husbands' occupations were extracted through record linkage with the Gibraltar Government's Births Registry and used as a proxy to assign individuals to high, medium, or low socioeconomic status (SES). We acknowledge that this study is constrained by its retrospective protocol and indirect evidence drawn from sociodemographic information. Written information on breastfeeding in Gibraltar is extremely limited and infant feeding is a subject matter that was seldom commented on directly, or in any great detail, by health officials in their annual reports.

This study examines trends in the incidence of breastfeeding, but does not track the duration of breastfeeding over infancy. Since women tend to switch from breastfeeding to artificial feeding, and not the other way around (Hirschman and Butler 1981), the reported rates are likely to be a maximum. We focus on feeding choices among primiparous women. It has been observed that feeding choices can differ significantly between primiparous and multiparous women, as women who had fed prior infants may be influenced by those individualized experiences (e.g., a failure to successfully breastfeed the first child may lead to a decision not to breastfeed subsequent children). Hirschman and Butler (1981) note that breastfeeding rates tend to be highest for first births. Had we included all women giving birth in Gibraltar in the period of time under study, breastfeeding rates would likely be lower. Our sample includes only Gibraltarian women (born in Gibraltar) in an effort to reduce heterogeneous cultural influences, experiences, and perceptions regarding infant feeding.

To remove any biologically derived confounding influences on feeding practices, the study is limited to infants who weighed at least 2500 grams and were full term at birth. Infant feeding choices were recorded in the maternity department summary register at the time of discharge. Consequently, our data concerns infant feeding behavior established in-hospital. Adjustments may have occurred once women returned to their homes. Our measure of 'breastfeeding' includes both exclusive breastfeeding and women breastfeeding with complementary feeds (of formula). The measure is intended to capture the general willingness to initiate breastfeeding (at least within the hospital setting).

We should note that there were changes in medical practice over the study period that could have implications for breastfeeding. Obstetric procedures, such as caesarean deliveries, have been found to contribute to a decreased likelihood of breastfeeding (Rajan 1994). Intuitively, any situation where either the mother or infant requires special care postpartum may influence breastfeeding. The relative impact of "childbirth technologies" (caesarean or forceps deliveries) varied over the study period. Changing obstetric trends may be reflecting the preferences of obstetricians and/or women in Gibraltar (and, of course, the availability of technologies to make these assisted deliveries possible). While forceps deliveries were more prominent in the 1970s, there was a marked increase in caesarian deliveries in the 1990s. Regression analyses on deliveries in this community between 1960–1992 revealed that caesarean deliveries, but not forceps deliveries, were associated with increased odds of artificial feeding (Burke 1999). As a result, we appreciate that the variable nature of medical practice in the community may have played a role in breastfeeding trends.

To undertake a quantitative examination of secular trends in breastfeeding behavior, a birth cohort approach was taken. Four temporal cohorts were reconstituted to capture the marked sociocultural, political, and economic conditions surrounding the border closure: 1955–1964 (open border), 1965–1974 (early closing/closed border), 1975–1984 (late closed border), and 1985–1994 (reopened border).

Methodologically, breastfeeding trends were analyzed using the proportional analytic approach outlined by Fleiss (1981: 138–59). The method provides information in terms of proportions of women breastfeeding their children for a given cohort or group. Statistically, the analysis employs a series of chi-square tests. If appropriate, results are partitioned into a series of further hypotheses regarding cohort differences and tests of group homogeneity (pooled cohorts). Since partitioning was suggested by the data, not planned a priori, the values of both difference and group are referred to the chi-square critical value where the degrees of freedom equals the number of cohorts minus one.

Results: Secular Trends in Infant Feeding in Twentieth-Century Gibraltar

Breastfeeding rates in Gibraltar reveal important changes in infant feeding over time. Table 5.1 presents the proportions of breastfeeding by cohort. From the first cohort (1955–1964) to the second cohort (1965–1974), the proportion of breastfeeding mothers fell by approximately 50 percent, a highly significant decline. While there was a significant increase in the proportion of breastfeeding between the second and third cohort (1975–1984), the difference was not as dramatic. Combining cohorts three (1975–1984) and four (1985–1996) proved not to be statistically significant, suggesting that over these two periods the proportion breastfeeding remained relatively stable.

Table 5.1. Proportion of infants breastfed in Gibraltar by cohort.

Cohort	Year	Proportion Breastfed	N
1	1955–64	.911	675
2	1965–74	.499	579
3	1975–84	.618	610
4	1985–96	.657	1002
Total		.657	2866

	Hypothesis tested	χ^2	df	Significance
Difference	Cohort 1 vs. Cohort 2	234.8	3	$p < .001$
Difference	Cohort 2 vs. Cohort 3	18.6	3	$p < .001$
Difference	Cohort 2 vs. Cohort (3+4)	21.1	3	$p < .001$
Group	Homogeneity (Cohort 3+4)	2.9	1	not significant
Total		275.21	3	$p < .001$

Overall, the incidence of breastfeeding in Gibraltar ranks much higher than observations in the United States at similar points in the twentieth century (see Jelliffe and Jelliffe 1978; Hirschman and Butler 1981). Gibraltar differs from other Western countries in that breastfeeding was nearly universal in the 1950s, while other countries had already begun to see a decline in breastfeeding rates. By the early 1970s, breastfeeding rates reached their lowest point, both in Gibraltar and other nations. Therefore, the steep decline in breastfeeding in Gibraltar in the 1960s represented, in effect, a "catching up" of Gibraltar to other industrialized nations. Both Gibraltar and the rest of the developed world have seen a resurgence of breastfeeding since the 1970s.

Using occupation as a proxy for SES, further analyses consider the effects of class differentials on breastfeeding behavior over time (Table 5.2). In the first cohort, 1955–1964, no significant class differences in infant feeding decisions are observed, with breastfeeding favored among women of all economic groups. The same is true for the second cohort, 1965–1974, when the decline in the choice to breastfeed occurred among all women, regardless of social class. Similarly, there are no class differentials in the third cohort, with the resurgence in breastfeeding being experienced by women in all socioeconomic groups. In contrast, significant class differences emerge during the last cohort (1985–1996), namely, those women giving birth to their first child after the border reopened. It is apparent that the discrepancy between the three classes lies principally in the rising rates of breastfeeding among the wealthy relative to the pooled women of middle and low SES. Therefore, SES is an important predictor of infant feeding choices only for the last cohort, implying that class differentials in infant feeding behavior is purely a recent phenomenon in Gibraltar.

Discussion: Gibraltar's Secular Trend in Infant Feeding

Pre–World War II

Written records on breastfeeding in Gibraltar prior to World War II are sparse but invaluable if we are to consider the issue of secular trends. Indirect information comes from the local newspaper, as advertisements for artificial formula were commonplace by 1910, if not earlier. Early medical perspectives come from the 1920 Annual Report on Public Health where it was reported that "in the great majority of cases infants have been breast-fed" (AR PH 1920: 41). The report also alludes to significant "propaganda work," including the distribution of four thousand leaflets, to educate the public on the importance of breastfeeding for infant wellness.

A rudimentary Child Welfare Centre (see Figure 5.2) opened in Gibraltar in 1918 to assist with monitoring infant health in the postpartum period. Mothers brought along the Centre's "Weight Card and Food Record" to each visit so that infant weight could be carefully recorded. The support for breastfeeding is evident, as the card advises mothers that "no form of artificial feeding is generally so safe or

Table 5.2. Proportion of infants breastfed in Gibraltar by socioeconomic status (SES) and cohort

	Proportion Breastfed	N	CV*
Cohort 1: 1955–1964			
High	.933	15	6.4
Middle	.851	174	2.7
Low	.902	41	4.6
Total	.865	230	
Cohort 2: 1965–1974			
High	.600	45	7.3
Middle	.514	403	2.5
Low	.452	115	4.6
Total	.508	563	
Cohort 3: 1975–1984			
High	.697	43	7.0
Middle	.617	434	2.0
Low	.629	124	4.3
Total	.625	601	
Cohort 4: 1985–1996			
High	.778	99	4.2
Middle	.600	585	2.0
Low	.542	277	3.0
Total	.601	961	

Hypothesis tested for Cohort 4		χ^2	df	Significance
Difference	High vs (Middle & Low)	14.31	2	$p < .001$
Group	Homogeneity (Middle & Low)	2.68	1	not significant
Total		16.99	2	$p < .001$

*CV is the coefficient of variation for the proportion breastfed. This parameter measures the relative amount of variability in the proportion of women who are breastfeeding within specified groups.

Figure 5.2. Interior of the Maternity and Child Welfare Centre: Gibraltar.

satisfactory as the mother's milk" and alludes to the problem of flies contaminating feeds of bottle-fed infants. The card is also revealing in that it succinctly links a focus on infant weight gain and anxieties over not producing enough breast milk. In this context, the attraction to supplemental feeds, or the altogether abandonment of breastfeeding, is evident.

The 1934 Annual Report on the Health of Gibraltar provides further insight into infant feeding practices in Gibraltar for a small number of women (n=108) in receipt of welfare support. In these cases:

> The Nestlé and Anglo-Swiss Condensed Milk Company has again supplied milk and certain infant foods, also feeders, at a reduced rate. ... this assistance is greatly appreciated as the external maternity, home visiting and welfare work depend on a grant from the Colonial Government assisted by the small contributions by mothers for milk, etc., supplied at the Welfare Centre. (AR Health of Gibraltar 1934: 13)

Cow's milk, Glaxo, and Virol were all available through the Centre and, though health authorities continued to promote breastfeeding, offered as alternatives to mothers who could not or would not breastfeed, or as supplementary feeds for older infants. By the 1940s, the Infant Welfare Centre had stepped up efforts to ensure the health of infants, offering Adexolin (vitamin A and D preparation), Cod Liver Oil (vitamin D), Fersolate (iron for anemia), Ostocalcium tablets for healthy bones, vitamin C tablets, and Gentian Violet (for thrush) (AR 1946: 8). Unfortunately, well into the 1960s it was found that poor and uneducated mothers most in need of this service were less likely to attend the clinic.

Breastfeeding at Its Height: Pre-Border Closure (1955–1964)

The 1950s and early 1960s saw its own social turmoil with Gibraltarian families reuniting in the aftermath of World War II and resettling into their lives. Breastfeeding rates among primiparous women were high in the 1950s, with some 91 percent of women undertaking at least some breastfeeding prior to hospital discharge. Gibraltar is somewhat anomalous in its high breastfeeding prevalence; comparable rates in the United States and Norway at this time suggested dramatic declines in breastfeeding, with only one-third to one-half of women breastfeeding (Liestøl, Rosenberg, and Walløe 1988; Ryan 1997). However, breastfeeding rates in Gibraltar were still declining in this period, just not to the same extent as other industrialized countries.

In the 1950s, a local physician, Dr. Triay, began to voice concern that not all Gibraltarian women were breastfeeding their infants. He attributed the decline to an "ignorance on the serious consequences which omission of breast feeding may have on the future nutrition and health of ... babies," women's own doubts in their breastfeeding abilities, and the perception that artificial foods were just as good as breast milk (Triay 1955: 45–46). The same reasons are widely given for declines in breastfeeding in other countries (Apple 1987; Hausman 2003; Lutter 2000; Van Esterik 1989).

It is interesting that breastfeeding in Gibraltar, while following similar patterns to other countries, lagged behind in the breastfeeding decline. Gibraltar, due to its unique geography and position as a colonial outpost, is effectively an island, isolated linguistically, culturally, and politically from neighboring Spain. However, the displacement of women during the Second World War opened the eyes of the population to the outside world in a way that had not been done before. As women returned to Gibraltar after the war, they may have begun to practice behaviors they had witnessed during their years abroad. Since breastfeeding had been already largely replaced by artificial feeding in many of the countries that harbored Gibraltarian women during the war, it is not surprising that women returning to Gibraltar might be disinclined to breastfeed. The net effect of the evacuation could have shifted the "innovator effect" typically associated with SES, as the uniform removal of all women from Gibraltar meant that all were privy to new approaches, observations, and experiences.

Breastfeeding on the Decline: The Border Closing/Early Border Closure (1965–1974)

Perhaps one of the most often-cited factors associated with the decline in breastfeeding is the rise in women's paid employment in the twentieth century (Dobbing 1988; Gengler, Mulvey, and Oglethorpe 1999; Hirschman and Butler 1981). In Gibraltar, women's participation in the labor force rose dramatically, largely in connection with the profound local changes driven by the border closure. Prior to World War II, it was very rare that Gibraltarian women sought employment outside the home (AR DLW 1956: 22). After repatriation, there was a growing acceptance of unmarried Gibraltarian women in private industry resulting in an increase from 16 to 44 percent employed between 1946 and 1956 (AR DLW 1956: 22). It was still Spanish women, however, who traveled across the frontier and accepted most of the available positions in domestic work, catering services, tobacco manufacturing, and dressmaking and tailoring (AR 1949: 8; AR DLW 1952/53: 20). Culturally, Gibraltarian women's employment was viewed as incidental, something to "do" in between leaving school and getting married.

New opportunities for, and official encouragement of, Gibraltarian women's employment grew out of Spanish-English tensions over Gibraltar. Though we will highlight a few salient points here, a more thorough discussion can be found in Burke and Sawchuk (2007). In the years leading up to the full border closure, the Spanish government enacted a number of restrictions on Spanish workers in Gibraltar. In the 1950s, the issuing of new work permits was discontinued so that only those holding previously issued permits could continue their employment on the Rock. In the 1960s, Gibraltar was also beginning to feel the effects of the wartime birth shortage and, paired with the declining number of Spanish workers, authorities appealed to Gibraltarian women to enter into employment.

In an effort to encourage women into the workforce, government-associated "Official Employers" attempted to accommodate some of the needs of women with families. Married women received special maternity leave arrangements, a provision not necessarily offered by private employers (AR 1964: 42). Paid maternity leave can affect breastfeeding decisions since women with no maternity benefits must return to work quickly, and are less likely to breastfeed (Escribà et al. 1994; Liestøl, Rosenberg, and Walløe 1988; Quandt 1995; Van Esterik 2002).

Government measures to attract women to the workforce led to satisfactory results up to 1964 (AR 1965), but then there was another shock. In 1965, all married Spanish women were required to surrender their frontier working passes, effectively removing them from the Gibraltarian workforce (AR 1965: 12). Then, in August 1966, *all* female Spanish frontier workers, some two thousand women, were denied their regular entry into Gibraltar by Spanish border guards (AR DLSS 1965/66: 14). A labor crisis ensued. Gibraltarian women responded, offering their voluntary services to keep vital institutions running; Martens (1987) conceptualizes these actions as an expression of patriotism and commitment to Gibraltar. While initially working on a voluntary basis, many women eventually took up permanent paid employment, finding they were able to cope with the extra demands and welcoming the increased household income (AR 1966: 5).

With the full closure of the border and the Manpower Mission Report of 1969, it was determined that a "largely untapped group" of several thousand married women were not working because they had young children. According to the report, "opinion [was] very divided about the social desirability of encouraging a higher rate of employment in this group, but, because of the compactness of the city, the effects upon family life of mothers going out to work need not be as great as they are in other situations" (PR no. 162/69 1969: 6). A new Government Nursery opened late in

1969, the first of its kind. Though innovative, the nursery could only accommodate twenty children in its first year of operation (AR 1969: 31). Nonetheless, the very construction of the nursery sent a clear message that the government was interested in supporting employment ambitions among Gibraltarian mothers.

According to one report from 1979, a rising trend towards young mothers returning to paid employment within two to three months following childbirth had been observed (AR DMHS 1979). When asked why they returned so soon, mothers responded that they needed the money or wanted to resume the pursuit of their careers (AR DMHS 1979: 25). Infants were typically left with their grandmothers. Sharing childcare with family members was also enhanced by the multigenerational households typical in Gibraltar until the 1990s, fueled, in part, by a lack of affordable housing. This aspect alone may have important implications since Mohrer (1979) found that young women living in parental homes were less likely to breastfeed since they were more likely to share childcare tasks with their mothers.

Alongside the changes occurring in the workplace, childrearing practices were also drawing attention. By 1973, declining interest in breastfeeding and a seemingly rising preference for chubby babies attracted concern from the local medical community:

> There is very little desire to breast feed babies, but there seems to be an overwhelming need in mothers to spoon feed their babies on soups and cereals at a very early age. There are many overweight babies—my department is consistently trying to educate mothers in the intricacies of a well-balanced diet. Mothers continue to give all kinds of vitamins to their children, we try to discourage this practice and tell mothers that vitamins should be prescribed by the child's doctor (AR MHD 1973).

In interviews conducted earlier by Burke (1999), it became apparent that the significance of infant weight gain was not underappreciated in the community. Mothers reportedly favored bottle feeding since they could monitor the quantity of food consumed by their infants, a level of precision not possible with breastfeeding. The perception, either real or imagined, that the infant is not receiving enough food can cause a strong emotional response in the mother (Heinig et al. 2006), and anxiety over whether her milk supply is sufficient can biologically interfere with a mother's ability to produce milk (Quandt 1995). According to Heinig and co-workers (2006), mothers generally tend to be concerned if a child is underweight, but do not mind or are even proud if their child is a bit heavy.

A Return of Breastfeeding: The Late Closure (1975–1984) and Border Reopening (1985–1996)

It is important to frame social changes within larger-scale events that can impact human behavior. In the case of Gibraltar, two events exerted significant economic impacts: the border closure and the hyperinflation associated with the great oil hikes of 1973 and 1979. These shifts resulted in a decade of economic instability when Gibraltar's economy was already troubled by the closing of the naval dockyard. Nonetheless, according to the government budget speech of 1984, financial repercussions for most Gibraltarians were held at bay through an overall increase in wages and a greater participation of women in the labor force. While it is impossible to directly attribute changing breastfeeding practices to the economy, it is possible that formula feeding became more affordable as the disposable family income rose.

The upswing in the rate of breastfeeding in the last two cohorts of the study correlates well with the general rise in breastfeeding observed in other Western industrialized countries of the time (see Hendershot 1984). Socially, however, these two cohorts are markedly different, with late 1970s–early 1980s Gibraltar still in the midst of the border closure, while 1980s–1990s Gibraltar saw an open border and a growing interconnectedness with Spain.

When the border reopened in 1985, Gibraltar changed dramatically. Unprecedented levels of tourism shot through the community, well beyond those observed before the border closed in the 1960s. More than ever before, Gibraltar was exposed to individuals and cultures from far and wide. As the exposure of women to other cultures during World War II had the potential to shape and change perceptions on infant feeding, so too did the increase in tourism and interconnectedness that followed the reopening of the border. However, while the fashion in the 1950s was to use artificial formula, this time Gibraltarian women were exposed to trends of increased breastfeeding. The economic boost instilled by tourism and government support meant that an increasing number of Gibraltarians could leave temporarily for extended education abroad (typically in the UK). On all levels, the community that emerged when the border reopened was increasingly cosmopolitan. This did not mean an easy or equal transition for all, however, since the open border meant the return of Spanish workers and employment became more competitive. Within this context of opportunity mixed with competition, socioeconomic stratification of Gibraltarian women and their families became more apparent, the effects of which potentially extended to infant feeding choices.

The Emergence of Class Differentials in Breastfeeding Rates

A recurrent theme in the literature is that women of high SES drive breastfeeding trends in communities depending on what is perceived as desirable (or even "fashionable") at the time (Liestøl, Rosenberg, and Walløe 1988; Lutter 2000; Van Esterik 2002). The findings from Gibraltar on social class and breastfeeding cannot directly address the issue of high SES as forerunners of social change. However, two salient points can be made about infant feeding and class.

First, in the three cohorts from 1955 through 1984 there were no statistically significant socioeconomic-related differences in breastfeeding practices. The data suggests that until the last cohort, Gibraltarian women responded as a collective whole in the broad ebb and flow of rates. Coefficients of variation in proportions of breastfeeding over the four cohorts are very low, never exceeding the 7 percent range (see Table 5.2). High SES women display the greatest levels of relative variability in proportion breastfeeding (ranging from 4.2 to 7.3 percent), while middle SES women show the smallest variability in proportion breastfeeding (ranging from 2.0 to 2.7 percent) (low SES is intermediate). The coefficient of variation suggests that the response to breastfeeding trends was relatively homogenous over time, with the greatest potential variability observed among high SES women. Given that infant feeding decisions are a highly complex aspect of human behavior, the uniformity of response is remarkable and suggests a shared communal response to complex factors over time.

Second, it is only in the last cohort (1985–1996) that significant differences with respect to SES emerge, with higher rates of breastfeeding observed among women of higher SES. As evident in Figure 5.3, there is a marked divergence in breastfeeding patterns with rising rates of breastfeeding rates among high SES women and declining rates among middle and low SES women, roughly corresponding with the border reopening. A retrospective study of this nature cannot address the factor(s) responsible for the observed divergence, but it is worth commenting that with the reopened border and greater access to information, educational differences emerging among women, and the widening of the economic gap between the rich and poor, what was once a community with shared values now sees fundamental differences emerging at its very core.

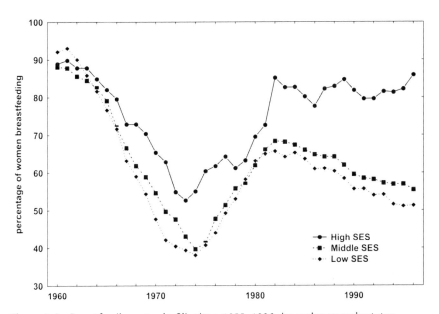

Figure 5.3. Breastfeeding rates in Gibraltar: 1955–1996, by socioeconomic status.

Conclusion

As a tightly knit population, Gibraltar experienced its own unique sociopolitical disturbances, including the evacuation/repatriation of World War II and the fifteen-year border closure with neighboring Spain. Yet, despite the singularity of these experiences, Gibraltarian women would eventually join other industrialized populations in the decline and return of breastfeeding. The perceived scientific value and convenience of artificial feeding infiltrated Gibraltar, as elsewhere, by the media, and by women who had lived abroad during the war years or spent time in British schools. Our research shows that there can be important local variations in breastfeeding. The effects of SES in relation to breastfeeding has captured significant attention, with women of high SES often pegged as driving forces in secular trends. A complex factor, SES is often tied to differences in wealth, education, access to information and support, and employment patterning in the postpartum period. Our research suggests that over much of the study period, high SES women were not innovators of secular change in infant feeding. Significant socioeconomic differences emerge only in the latest cohort since 1985, as high SES women departed significantly from either middle- or low-SES groups in favoring higher levels of breastfeeding.

Acknowledgements

We wish to thank Dennis Beiso (Gibraltar Government Archives) and Clive Finlayson (The Gibraltar Museum) for assistance in locating images and Ken Jones (University of Toronto at Scarborough) for technical expertise in clarifying the images for publication. This research was supported by a grant from the Social Sciences and Humanities Research Council of Canada.

References

Abstract of Statistics 1995. Government of Gibraltar.
Annual Report on Gibraltar 1946. Government of Gibraltar.
Annual Report on Gibraltar 1949. Government of Gibraltar.
Annual Report on Gibraltar 1964. Government of Gibraltar.
Annual Report on Gibraltar 1965. Government of Gibraltar.
Annual Report on Gibraltar 1966. Government of Gibraltar.
Annual Report on Gibraltar 1969. Government of Gibraltar.
Annual Report on the Health of Gibraltar 1934. Government of Gibraltar.
Annual Report on the Health of Gibraltar 1959. Government of Gibraltar.
Annual Report, Department of Labour and Social Services 1965/66. Government of Gibraltar.
Annual Report, Department of Labour and Welfare 1952/53. Government of Gibraltar.
Annual Report, Department of Labour and Welfare 1956. Government of Gibraltar.
Annual Report, Department of Medical and Health Services 1979. Government of Gibraltar.
Annual Report, Medical and Health Department 1973. Government of Gibraltar.

Annual Report on Public Health 1920. Government of Gibraltar.
Annual Report on Public Health 1926. Government of Gibraltar.
Apple, R. D. 1987. *Mothers and Medicine: A Social History of Infant Feeding, 1890–1950*. Madison: University of Wisconsin Press.
———. 1994. The medicalization of infant feeding in the United States and New Zealand: Two countries, one experience. *Journal of Human Lactation* 10: 31–37.
Bauchner, H., J. M. Leventhal, and E. D. Shapiro. 1986. Studies of breast-feeding and infections: How good is the evidence? *Journal of the American Medical Association* 256: 887–92.
Burke, S. D. A. 1999. "A home, a husband, and now a baby: The implications of premarital conception in Gibraltar, 1960–1996." PhD thesis, Department of Anthropology, University of Toronto.
Burke, S. D. A., and L. A. Sawchuk. 2007. Reproductive choices in Gibraltar: A case study of a community in transition, 1960–1996. *Canadian Studies in Population*: 34 (2): 149–78.
Cunningham, A. S. 1995. Breastfeeding: Adaptive behavior for child health and longevity. In *Breastfeeding: Biocultural Perspectives*, ed. P. Stuart-Macadam and K. A. Dettwyler, 243–64. New York: Walter de Gruyter, Inc.
Dettwyler, K. A., and C. Fishman. 1992. Infant feeding practices and growth. *Annual Review of Anthropology* 21: 171–204.
Dobbing, J. 1988. Medical and scientific commentary on charges made against the infant food industry. In *Infant Feeding: Anatomy of a Controversy 1973–1984*, ed. J. Dobbing, 9–28. London: Springer-Verlag.
Escribà, V., C. Colomer, R. Mas, and R. Grifol. 1994. Working conditions and the decision to breastfeed in Spain. *Health Promotion International* 9: 251–58.
Fildes, V. 1986. *Breasts, Bottles, and Babies: A History of Infant Feeding*. Edinburgh: Edinburgh University Press.
Finlayson, T. J. 1996. *The Fortress Came First: The Story of the Civilian Population of Gibraltar During the Second World War*. Grendon, UK: Gibraltar Books Ltd.
Fleiss, J. L. 1981. *Statistical Methods for Rates and Proportions*. 2nd ed. New York: John Wiley & Sons.
Gengler, C. E., M. S. Mulvey, and J. E. Oglethorpe. 1999. A means-end analysis of mothers' infant feeding choices. *Journal of Public Policy & Marketing* 18: 172–88.
Gold, P. 1994. *A Stone in Spain's Shoe: The Search for a Solution to the Problem of Gibraltar*. Liverpool: Liverpool University Press.
Government of Gibraltar 1970. Census of Gibraltar.
Government of Gibraltar 1981. Census of Gibraltar.
Government of Gibraltar 1991. Census of Gibraltar.
Government of Gibraltar 2001. Census of Gibraltar.
Hausman, B. L. 2003. *Mother's Milk: Breastfeeding Controversies in American Culture*. New York: Routledge.
Heinig, M. J., J. R. Follett, K. D. Ishii, K. Kavanagh-Prochaska et al. 2006. Barriers to compliance with infant-feeding recommendations among low-income women. *Journal of Human Lactation* 22: 27–38.

Hendershot, G. E. 1984. Trends in breast-feeding. *Pediatrics* 74: 591–602.
Hirschman, C., and M. Butler 1981. Trends and differentials in breast feeding: An update. *Demography* 18: 39–54.
Hofvander, Y. 2005. Breastfeeding and the baby friendly hospitals initiative (BFHI): Organization, response and outcome in Sweden and other countries. *Acta Paediatrica* 94: 1012–16.
Holman, D. J., and M. A. Grimes. 2003. Patterns for the initiation of breastfeeding in humans. *American Journal of Human Biology* 15: 765–80.
Horta, B. L., C. G. Victora, D. P. Gigante, J. Santos et al. 2007. Breastfeeding duration in two generations. *Rev Saúde Pública* 41: 1–5.
Jackson, W. G. F. 1987. *The Rock of the Gibraltarians: A History of Gibraltar.* Grendon, UK: Gibraltar Books Ltd.
Jelliffe, D. B., and E. F. P. Jelliffe. 1978. *Human Milk in the Modern World: Psychosocial, Nutritional, and Economic Significance.* Oxford: Oxford University Press.
Kintner, H. J. 1985. Trends and regional differences in breastfeeding in Germany from 1871 to 1937. *Journal of Family History* 10: 163–82.
Knaak, S. 2005. Breast-feeding, bottle-feeding and Dr. Spock: The shifting context of choice. *The Canadian Review of Sociology and Anthropology* 42: 197–216.
Kramer, M. S., and R. Kakuma. 2002. *The Optimal Duration of Exclusive Breastfeeding: A Systematic Review.* Geneva: World Health Organization.
Lanting, C. I., J. P. Van Wouwe, and S. A. Reijneveld. 2005. Infant milk feeding practices in the Netherlands and associated factors. *Acta Paediatrica* 94: 935–42.
Liestøl, K., M. Rosenberg, and L. Walløe. 1988. Breastfeeding practice in Norway 1860–1984. *Journal of Biosocial Science* 20: 45–58.
Losch, M., C. I. Dungy, D. Russell, and L. B. Dusdieker. 1995. Impact of attitudes on maternal decisions regarding infant feeding. *The Journal of Pediatrics* 126 (4): 507–14.
Lutter, C. K. 2000. Breastfeeding promotion—is its effectiveness supported by scientific evidence and global changes in breastfeeding behaviours? In *Advances in Experimental Medicine and Biology, Vol 478: Short and Long Term Effects of Breast Feeding on Child Health,* ed. B. Koletzko, K. Fleischer Michaelsen, and O. Hernell, 355–68. New York: Kluwer Academic Publishers.
Martens, J. L. 1987. "Gibraltar and the Gibraltarians." PhD thesis, University of London.
Martinez, G. A., and J. P. Nalezienski. 1981. 1980 update: The recent trend in breast-feeding. *Pediatrics* 67: 260–63.
McNally, E., S. Henericks, and I. Horowitz. 1985. A look at breast-feeding trends in Canada 1963–1982). *Canadian Journal of Public Health* 76: 101–7.
Mohrer, J. 1979. Breast and bottle feeding in an inner-city community: An assessment of perceptions and practices. *Medical Anthropology* 3: 125–45.
Press Release (PR) no. 166/67. 1967. Gibraltar Government Archives.
Press Release (PR) no. 162/69. 1969. Gibraltar Government Archives.
Press Release (PR) 30 January 1981. Gibraltar Government Archives.

Quandt, S. A. 1995. Sociocultural aspects of the lactation process. In *Breastfeeding: Biocultural Perspectives*, ed. P. Stuart-Macadam and K. A. Dettwyler, 127–44. New York: Walter de Gruyter, Inc.

———. 1998. Ecology of breastfeeding in the United States: An applied perspective. *American Journal of Human Biology* 10: 221–28.

Rajan, L. 1994. The impact of obstetric procedures and analgesia/anaesthesia during labour and delivery on breast feeding. *Midwifery* 10: 87–103.

Ryan, A. S. 1997. The resurgence of breastfeeding in the United States. *Pediatrics* 99: E12–E16.

Ryan, A. S., Z. Wenjun, and A. Acosta. 2002. Breastfeeding continues to increase into the new millennium. *Pediatrics* 110: 1103–09.

Sawchuk, L. A. 1992. Historical intervention, tradition, and change: A study of the age at marriage in Gibraltar, 1909–1983. *Journal of Family History* 17: 69–94.

Sawchuk, L. A., S. D. A. Burke, and S. Benady. 1997. Assessing the impact of adolescent pregnancy and the premarital conception stress complex on birthweight among young mothers in Gibraltar's civilian community. *Journal of Adolescent Health* 21: 259–66.

Scott, J. A., M. C. G. Landers, R. M. Hughes, and C. W. Binns. 2001. Psychosocial factors associated with the abandonment of breastfeeding prior to hospital discharge. *Journal of Human Lactation* 17: 24–30.

Shorter, E. 1985. *Doctors and Their Patients: A Social History.* New Brunswick, NJ: Transaction Publishers.

Singh, G. K., M. D. Kogan, and D. L. Dee. 2007. Nativity/immigrant status, race/ethnicity, and socioeconomic determinants of breastfeeding initiation and duration in the United States, 2003. *Pediatrics* 119: S38–S46.

Siskind, V., C. Del Mar, and F. Schofield. 1993. Infant feeding in Queensland, Australia: Long term trends. *American Journal of Public Health* 83: 103–5

Stewart, J. D. 1967. *Gibraltar, the Keystone.* London: John Murray.

Tellis, W. 1997. Introduction to case study. *The Qualitative Report* 3 (2) http://www.nova.edu:80/ssss/QR/QR3-2/tellis1.html.

Triay, H. G. 1955. "History of Infant Medical Care in Gibraltar." MA thesis, Department of Medicine, University of Aberdeen. Gibraltar Government Archives.

Tönz, O. 2000. Breastfeeding in Modern and Ancient Times: Facts, Ideas and Beliefs. In *Advances in Experimental Medicine and Biology,* vol. 478, *Short and Long Term Effects of Breast Feeding on Child Health,* ed. B. Koletzko, K. Fleischer Michaelsen, and O. Hernell, 1–21. New York: Kluwer Academic Publishers.

Van Esterik, P. 1989. *Beyond the Breast-Bottle Controversy.* New Brunswick, NJ: Rutgers University Press.

———. 2002. Contemporary trends in infant feeding research. *Annual Review of Anthropology* 31: 257–78.

World Health Organization 2003. *Global Strategy for Infant and Young Child Feeding.* Geneva: World Health Organization.

*Food Insecurity
and Malnutrition*

• 6 •
Dietary Diversity, Dietary Transitions, and Childhood Nutrition in Nepal

Questions of Methodology and Practice

T. Moffat and E. Finnis

Introduction

In this chapter, we examine issues of dietary diversity and food transitions in Nepal. In particular, we ask how the move from a rural to urban environment might influence the diversity of children's diets. Although there are numerous circumstances under which rural Nepalis might move to urban settings and urban forms of employment, we focus specifically on the diets of the children of economically marginalized carpet factory workers. In doing so, our discussion is both substantive and speculative. We engage with Dietary Diversity Scores (DDS) as part of a methodology to assess nutrition, food practices, and the practicalities of dietary transitions. We address issues of chronic undernutrition and suggest that current dietary transition research focusing on overnutrition and obesity in developing countries risks downplaying questions of how dietary transitions might contribute to nutritional insufficiency. In the final part of the chapter, we turn to a more speculative discussion of dietary transitions and the moral economies of food. We suggest that qualitative methodologies that draw on local voices about food values and preferences need to be included in research on the health implications of food regime changes and dietary transitions.

Analyses of dietary transitions and nutritional shifts have been approached from a variety of perspectives, both within anthropology, and in the food and nutrition sciences. Archaeological and evolutionary approaches, for example, examine both subtle and larger-scale dietary transitions and the methodologies that can be employed to assess these changes (e.g., Molleson, Jones, and Jones 1993; Leonard 1994; Cachel 1997; Schmidt 2001). Analyses of changing agricultural regimes and settlement patterns, particularly in developing countries during the 1960s to the 1980s, spurred research on nutritional health and food production strategies of small farmers and foragers (e.g., Fleuret and Fleuret 1980; Dewey 1981; Dewey 1989; Shack, Grivetti, and Dewey 1990; Leonard 1992; Leonard et al. 1993; Leatherman 1994; Smith and Smith 1999; Kedia 2004).

More recent analyses of contemporary diet and nutrition shifts in developing countries have been increasingly concerned with one type of dietary transition: the shift towards a more "Western," modern diet that is characterized by high levels of fat, sugar, and processed, prepackaged food. Some of this research offers policy suggestions for nutrition professionals and governments in order to facilitate the promotion of a diet that is less dependent on highly processed foods (e.g., Popkin 1993, 1994; Haddad 2003). Other studies have considered the economic and human health costs associated with a modernizing diet that minimizes the consumption of foods containing high levels of fiber and antioxidants, while maximizing the consumption of meat products, sugars, and edible oils (e.g., Hijazi, Bahaa, and Seaton 2000; Popkin et al. 2001; Smil 2002). Much of this research explicitly addresses overnutrition and the relationship between obesity and health issues like diabetes and high blood pressure that are affecting, for example, adult Sherpa rural-to-urban migrants in Nepal (Smith 1998, 1999). This obesity and overconsumption research trend can be traced to a number of factors, including similar concerns with obesity in industrialized nations (see Katzmarzyk, this volume). As a "Western" diet has increasingly contributed to high or rising rates of obesity in North America, the UK, and Europe, there has been a corresponding interest in the effects of these globalized dietary practices in developing countries. Moreover, increased patterns of urbanization in developing countries, fewer incidences of severe famine around the world, and rapidly rising rates of Type-2 diabetes in countries like China and India have all helped to divert nutritional research away from questions of both acute and chronic undernutrition, particularly in South and Southeast Asia. While it may be acknowledged that undernutrition can exist side-by-side with overnutrition (Popkin 2001; Raphael and Delisle 2005), questions of chronic undernutrition may be downplayed.

In addition to focusing on issues of overnutrition, this work is often overwhelmingly macroscopic in scope, investigating broad trends in national, regional, or global contexts. There are few examinations of why and how local communities make food choices. Haddad (2003), among others, has noted that one of the drivers of dietary and nutrition trends is the socioeconomic and activity changes accompanying urbanization. He is referring to broad processes, that is, global trends concerning dietary delocalization—in which populations increasingly rely on market sources of food that may have been grown, processed, and/or packaged in distant locales (Pelto and Pelto 1983)—and the shift to a "modern" diet. It is important to note, however, that these dietary and economic processes are interwoven with changes taking place at the local level that are shaped by, among other factors, specific economic circumstances (see Finnis 2007).

It is the specifics of one such trend that we explore in this chapter. We examine issues of dietary diversity and urbanization in Nepal, asking what Dietary Diversity Scores (DDS) can tell us about dietary patterns among Nepali children. We suggest that current nutritional research risks overlooking an important aspect of dietary diversity and urbanization, that is, the potential for chronic undernutrition that can accompany dietary transitions to market and preprocessed foods. Moreover, dietary transitions encompass more than questions of intakes. The process of dietary delocalization is intertwined with experiential aspects of food consumption, including food beliefs and preferences. More attention to these factors, we suggest, provides an emerging research agenda to better integrate quantitative analysis with the qualitative factors that shape and reflect food consumption and dietary transitions.

We approach this chapter with two case studies. The first examines issues of dietary diversity in a rural, high-altitude village setting. It demonstrates the value of locally raised foods in the diet of villagers. Prepared, commercial foods are not readily available in this rural setting. The second, larger case study examines the dietary intakes and nutritional health of the children of carpet factory workers in Kathmandu. These workers have left their rural communities in order to take advantage of the economic opportunities offered by the urban economy. In Kathmandu, people have easy access to a number of prepared, preprocessed foods that are not readily available in villages. At the same time, they experience processes of dietary delocalization and cannot afford to buy the range of fresh foods that may be more readily available in a rural setting where subsistence farming is the primary economic strategy.

Study Setting

The Kingdom of Nepal is a small Himalayan nation nestled between Tibet (People's Republic of China) and India. Its physical geography is diverse and can be partitioned into three parts: the plains, the Himalayan foothills, and the Himalayan mountain range. This study compares two locations, the rural setting in the foothills at 2000 m and the urban setting in the Kathmandu Valley at 1324 m above sea level.

The rural case study setting is a small village of about forty households in the Helambu region of Nepal in the district of Sindhupalchok, located in the Himalayan foothills north of Kathmandu. Though the village is not far in terms of distance from Kathmandu, the road goes only as far as the town of Melamchi Bazar, from where all traffic to the village is on foot up 1200 m. The village consists mainly of subsistence farmers. Services include a small primary school, and several small shops with limited foodstuffs for purchase such as biscuits, instant noodles, and some fruit. There are no healthcare facilities in the immediate vicinity. There is a trekking lodge in the village, but because it was not on the trekking circuit at the time of this study, it did not have much tourist business and was mostly used as a local restaurant. Most of the cash income entering the village comes from family members who have moved to Kathmandu or to foreign countries to work as migrant laborers. Many villagers had family members working in the carpetmaking industry in Kathmandu due to a connection with one village man who employed many of the young people in his factory. As well, several of the young girls had traveled to Kuwait to work as domestic workers and were sending remittances home.

Nepal has avoided the teeming megacities of its bigger neighbors in South Asia such as India, Pakistan, and Bangladesh. Nevertheless, the average annual growth rate of the urban population was 6.65 percent in the 1990s versus the national population growth rate of 2.25 percent. At this rate, half of the population of Nepal will be living in urban areas by the year 2035 (ADB 2000). Urbanization is just beginning in Nepal, but it is a rapid transition. Part of the impetus for it has been the movement of people from the rural villages to the cities to partake in the urban cash economy. With development and modernization goals in the minds of many Nepali people, moving to the city is viewed as an opportunity to gain educational and economic advancement. With this rapid change in human habitat, however, poverty has become more visible in urban Nepal, and the dependence on a market economy creates more inequities between the haves and have-nots.

The setting for the urban case study is periurban Kathmandu located approximately 10 km east of downtown Kathmandu. The area has a mix of rural and urban elements with the result that it is possible to see shops and industries adjacent to rice fields. The area is now connected by roads and public transportation to downtown Kathmandu, but it is still geographically, politically, and economically marginal from the city center. Many rural-to-urban migrants are attracted to this periurban location due to the plethora of carpet factories that provide wage labor opportunities and housing. However, in low-income countries worldwide, periurban centers are increasingly recognized as places replete with environmental and social problems (Harpham, Lusty, and Vaughan 1988). One major difficulty associated with living in periurban Nepal is the dearth of services and infrastructure. In particular, roads, solid-waste disposal, and water supply are inferior to such services found in downtown Kathmandu. Most families in this study do not have access to running water or latrines within their households, although most carpet factories provide public latrines.

The samples from both study settings are comprised of children from birth to five years. The urban sample consists of children whose mothers were working in some capacity in the carpet-making industry in periurban Kathmandu. Almost all of the women (95 percent) were born outside of Kathmandu, the majority originating from central and eastern Nepal. This is due both to the region's proximity to Kathmandu and to the fact that the hill and mountain regions are overpopulated relative to agricultural production; consequently there is a long tradition of out-migration for wage employment (Fricke 1993). Eighty-five percent of the migrant women had come to Kathmandu within the previous ten years and 57 percent had migrated within the previous five. Thus, the majority migrated during adolescence and early adulthood and arrived in Kathmandu either with their husbands or with family and friends. Ethnically, they are diverse, representing all major ethnic groups from the central and eastern regions, with a preponderance of Parbatiya (hill Brahmins and Chetri), part of the Hindu majority, and Sherpa and Tamang peoples, part of the Buddhist minority.

The rural sample consists of all of the children less than five years of age residing in the village at the time of study. Many families had moved to Kathmandu or other countries in South Asia as migrant workers. Some settled in these places or had since returned to their village. The village is comprised of people who call themselves *Lamas* and are members of the ethnic group Tamang; one cluster of families who live higher up on a ridge above the main village call themselves *Yolmo*, or Helambu Sherpa (O'Neill 2001).

Seasonal changes in Nepal are marked by changes in both climate and agricultural production. In terms of climate, there are three distinct seasons: cold (October to March), hot (April to June), and rainy (July to September). The main agricultural season begins in April with the tilling of the soil and the planting of cereal crops. Rice paddies are planted at the beginning of the monsoon season and are harvested in the early fall. There are also winter crops at lower altitudes in the Kathmandu Valley and the warmer regions of the south.

The carpet workers in the city are almost completely reliant on food purchased from the market, although there are a few families that have small garden plots. As such, one might hypothesize that seasonality in diet is more likely to be linked with seasonal cycles in the carpet-making industry. The monsoon months are known as the "low season," when there are fewer orders from European buyers. Many workers remain at the factory, either idle or weaving stock carpets; some workers return to their home villages to fulfill family obligations.

The dietary data presented in the periurban case study is part of a larger study conducted by Moffat that documents Nepali children's growth, dietary, and infectious disease patterns. On average, the sample is characterized as being moderately to severely growth stunted[1], indicating chronic undernutrition. Indeed, 50.3 percent of the children from 6 to 36 months and 77.5 percent of children aged 36 to 60 months were growth stunted. Based on longitudinal growth data that was collected over a period of nine months, it is clear that growth stunting begins at around six months postpartum, the point at which infants are transitioning from exclusive breastfeeding to the introduction of complementary foods (Moffat 2001, 2002). No anthropometric data were collected for the village sample. However, researchers working in other rural villages in the Himalayan foothills and mountains report a high prevalence of growth stunting (Costello 1989; Huijibers et al. 1996; Panter-Brick 1997). In a comparison of boys living in a rural village in the Himalayan foothills and those living in a squatter community in Kathmandu, Panter-Brick, Todd, and Baker (1996) found that the urban children were, on average, slightly less stunted. It is important to note, though, that growth stunting is not just a result of poor nutrition; disease, particularly gastrointestinal diseases, and access to healthcare also play crucial roles in children's growth faltering (Martorell and Ho 1984; Lunn 2000; Moffat 2003; Panter-Brick et al. 2008).

Methods

Rural dietary data were collected over a three-week period in November 1998 with the assistance of a Nepali woman who lived in Kathmandu. She considered the village to be her ancestral home, as her parents were both born there. In periurban Kathmandu, the data were collected three times for each child from January to September 1995. In order to have the seasons correspond as much as possible between the rural and urban samples, only cross-sectional data from the cold season (Jan to Mar) was used for the urban sample.

To assess the variety of food that was consumed by the children in the study, a twenty-four-hour dietary recall survey (Gibson 1990: 37–39) was conducted with primary caregivers, usually the mother. In pilot tests of the recall, Moffat found that many mothers had difficulty remembering what their child had eaten. Thus, the procedure was modified by creating a list based on the types of food that participants had mentioned and on ethnographic knowledge of Nepali foods. The food frequency list was read to each mother and she then answered whether her child had eaten the item during the previous day. There is some loss of data when a list is used, since any food not on the list may go unreported. The list, however, produced a far more comprehensive summary of the diet than was possible to gather from a straight recall. Since most mothers were unable to specify the portions fed to their children and respondent burden was a concern, the quantity of the food consumed was not estimated. Therefore, the results of the survey can only be presented like those of a food frequency questionnaire.

Throughout the fieldwork in both locations, Moffat engaged in participant observation by visiting families in order to interact and observe women and children in their homes. This allowed for an understanding of the context of the food recall data.

Dietary Diversity

Dietary diversity, defined as the "number of individual foods or food groups consumed over a given time" (Hoddinott and Yohannes 2002: iii), is an important aspect of nutritional sufficiency and food security. A more diverse diet is one that provides better access to a variety of macro- and micronutrients, and consequently it has been argued that dietary diversity can provide one way to assess or triangulate food security and nutritional status, particularly from the perspective of the quality rather than quantity of the diet (Messer 1984; Hatloy, Torheim, and Oshaug 1998; Maunder, Matji, and Hlatshwayo-Molea 2001; Ruel 2003; Foote et al. 2004; Savy et al. 2005; Kennedy et al. 2007). Hoddinott and Yohannes have suggested that a less-diverse diet is more likely to consist of fewer calories.[2] Moreover, a varied diet, both between and within food groups, is associated with a number of potential health benefits including higher birth weights and lower risks of developing cardiovascular problems, hypertension, and some forms of cancer (Hoddinott and Yohannes 2002).

Dietary diversity is typically measured with the Dietary Diversity Score (DDS), which is defined as the number of different food groups consumed. The composition of these food groups can vary according to different cultural contexts. While some have used an understanding of food groups that correspond to North American norms (fruits, vegetables, grains, meat and alternate sources of protein, and dairy), more recently there have been attempts to standardize the DDS using ten food groups agreed upon during a 2004 FAO workshop on validation methods for dietary diversity held in Rome (outlined in Kennedy et al. 2007). This scoring method has subsequently been validated as indicators of nutrient adequacy by Steyn et al. (2006) for South African children aged one to eight years, and for micronutrient adequacy by Kennedy et al. (2007) for nonbreastfeeding children aged 24 to 71 months. Table 6.1 presents the ten standard food groups.

In our analysis only children over one year are included in the sample, as prior to one year of age many infants' diets are not very diverse because they are relying primarily on breastmilk, or in the case of a few, a breastmilk alternative, as a main source of nutrition. Since a large proportion of children between one and five years of age in

Table 6.1. Ten standard food groups for calculating Dietary Diversity Score (DDS). Adapted from Kennedy et al. (2007: 473).

Cereals and tubers
Meat, Poultry, and Fish
Dairy
Eggs
Pulses and Nuts
Vitamin A rich fruits and vegetables
Other fruit
Other vegetables
Oils and fats
Other (including sugar, non–juice or dairy beverages, and condiments and spices)

the samples were breastfeeding (61.9 percent of the urban sample and 41.7 percent of the rural sample), we did not exclude breastfeeding children, as done by Kennedy et al. (2007). Instead, we simply counted breastmilk as a dairy product food group. Like Kennedy et al. (2007) we did not count "other foods," such as sweets, as a food group, as they are not nutritionally beneficial.

Means and standard deviations were calculated for the DDS; however, we did not test the statistical difference between the mean scores for the rural and urban samples, as there are too few individuals in the rural sample. We did, however, compare the distribution of the scores for each sample.

Results

Rural Case Study

Figure 6.1 presents twenty-four-hour recalls of food frequencies for children aged birth to five years living in the village. All children under one year of age (n=6) were breastfeeding, but 14 of 24 children over one year were completely weaned. None of the infants were fed commercial breastmilk substitutes such as Lactogen or prepared weaning cereals such as Cerelac. These are included in Figure 6.3 for urban-rural comparison only. The first weaning cereal used in the village is a homemade rice porridge called *litho*.

There are several salient features of the village diet. First, for older children and adults, barley porridge is a staple food. This is not surprising as barley is an important high-altitude cereal crop, whereas rice cannot be grown and must be brought up from the lowlands where it can be cultivated. "Bread" refers to homemade *roti* (a flat

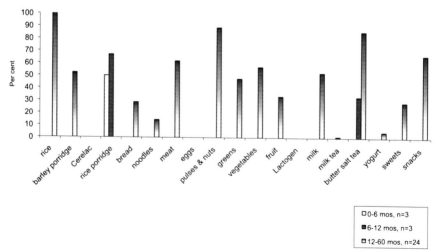

Figure 6.1. Food type consumed on recall day by rural children by age group.

bread made from flour). Given the lack of a road, it is not surprising that commercial bread and noodles are eaten in small quantities, although 20 percent of children were eating commercial biscuits that are brought up from the lowlands on shopping trips, or purchased at the small village shop. Second, meat was consumed by more than half of the children in the sample during the recall. It may be that it was anomalous, as someone may have recently killed a water buffalo, or goat, which would then be distributed or sold throughout the village; it should be noted, however, that meat is typically consumed at celebrations, and there was no such celebration occurring at the time of this study. Third, 80 percent of the children on the day of recall drank *nun-chiya*, salt butter tea, which is made by churning black tea, water buffalo butter, salt, and hot water. High-altitude villagers, including young children, consume this tea daily and in copious amounts.

The mean dietary diversity score (DDS) for the village children was 5.04, with a range of 2 to 7 and a standard deviation of 1.52. A histogram of the scores is shown in Figure 6.3. There are fewer scores at the lower end of the range, with over half of the village children scoring from 5 to 7.

Urban Case Study

Figure 6.2 presents food frequency data for the urban sample of children. In contrast to the rural sample, some of the children under one year were consuming Lactogen, a commercial breastmilk substitute, and Cerelac, a prepared weaning cereal. However, all of the infants under one year of age, except for two individuals, were still breastfeeding.

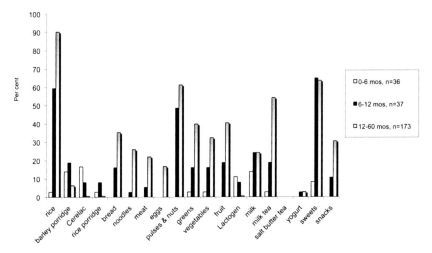

Figure 6.2. Food type consumed on recall day for urban sample by age group.

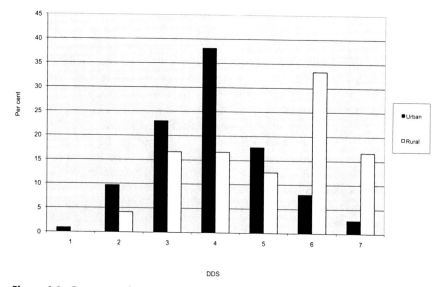

Figure 6.3. Frequency of Dietary Diversity Score (DDS) for rural and urban samples.

In contrast to the rural sample, the following features of the urban diet are noteworthy. First, rice is the main staple cereal; very few children ate barley porridge. This stands to reason, as rice is readily available for purchase in the city. It should be noted, though, that several children were fed barley cereal, a preferred food item by high-altitude villagers who have migrated to the city. Much of the bread consumed in the city was in the form of commercial white bread, although sometimes mothers made and fed their children *roti*. More urban children consumed commercial bread, noodles, and sweets (biscuits and puff pastry), compared to the village children. Second, noticeably fewer children ate meat on the day of recall compared to the rural children. There is no access to meat in the city except by purchasing it from a butcher shop. Meat is expensive and is only eaten by those with surplus cash to spend, and is thus usually reserved for important occasions. Third, none of the children drank salt butter tea in the city, despite the fact that there were many families who had migrated from high-altitude villages. It is possible that it is too labor intensive to churn the tea on a daily basis, or too expensive to buy the butter; thus it was not consumed, at least not by children. In contrast, many children in the city drank sweetened black tea that sometimes had purchased cow's milk added to it.

The mean DDS for the urban children is 3.96 with a range of 1 to 7 and a standard deviation of 1.20 (Fig. 6.3). These scores are normally distributed with the most frequent at the mean and very few at either end of the range.

Table 6.2 summarizes the DDS for the rural and urban samples. While we are unable to statistically test the mean difference between the scores due to the small

Table 6.2. Comparison of DDS for urban and rural samples.

Location	Sample Size	Mean DDS	STD DEV	Range
Rural	24	5.04	1.52	2 to 7
Urban	113	3.96	1.20	1 to 7

rural sample size, we can note the differences in the distribution of the scores and speculate on why they predominate at the higher end for the rural sample. Two foods that are definitely contributing to higher scores for rural children are the consumption of meat and salt butter tea. The high consumption of these foods by village children results in the addition of the poultry, fish and meat, and the oils and fat categories.[3] As well, more rural children drank milk readily available from water buffaloes and goats in the village. Urban children were much less likely to drink milk, which came in the form of cow's milk purchased from the national dairy corporation.

Discussion

Dietary Diversity and Nutritional Status

The dietary diversity data provide an analysis of the quality of dietary intakes in rural and urban settings, and can serve as an estimate of nutritional sufficiency. In particular, dietary diversity is an indicator of access to micronutrients, which are crucial for children under five years of age who are growing and developing rapidly and who are prone to infectious diarrheal and respiratory diseases. In particular, iron, zinc, and vitamin A deficiencies among others have been implicated in impaired growth and development deficits (Allen 1994; Diaz, de las Cagigas, and Rodriguez 2003). Though we cannot compare these two case studies statistically, due to the small sample size of the rural group, it appears from the distribution of the DDS for each group that the rural children are consuming a more diverse diet.

One limitation, however, is that these are cross-sectional surveys of the diets taken during the cold season. Seasonality does not have much impact on the urban sample; indeed an examination of mean DDS by season showed no statistically significant differences for the urban sample. Seasonal changes in diets, however, are well documented in rural Nepali communities (Costello 1989; Naborro et al. 1988; Panter-Brick 1997). This village sample was studied in November, one of the most bountiful points during the agricultural calendar, as it is postharvest. Thus, this is no doubt a time when children's diets are optimally diverse. Nevertheless, the contrast with the urban diet underscores the paucity of urban food diversity, particularly demonstrating that meat, dairy products, and fats and oils are minimal in the urban children's diets.

Though there is keen interest among nutritionists to use DDS as both a rapid assessment field tool and survey instrument, there is still much discussion of how to make it consistent and whether it is an appropriate indicator of nutritional sufficiency in all settings around the world (Ruel 2003; Arimond and Ruel 2004; Steyn et al. 2006). It appears that, for the most part, dietary diversity is positively correlated with macro- and micronutrient sufficiency, anthropometrics, and household food security, although less work has been done in this latter area (see Ruel 2003 for a review). There has also been improvement in making the DDS consistent for nine food groups (ten including "other foods," which is not counted in the score); these food groups were appropriate for the diet of the Nepali children in both the rural and urban case studies.

However, there are still some issues that remain unresolved. First, breastfed children have not been included in the studies that have validated the DDS (Steyn et al. 2006; Kennedy et al. 2007). Indeed, it appears that dietary diversity may not be as crucial for children who are still breastfed (Ruel 2003). This, however, makes it difficult to use the DDS for children in a country like Nepal, where the majority of children under five years continue to breastfeed (Moffat 2001, 2002). In our case, we decided to count breastmilk as part of the dairy products, but this deviates from the standard protocol outlined by Kennedy et al. (2007).

A second issue is the use of the DDS among groups that have very monotonous diets, for example, a vegetable-based diet. In that case a Food Variety Score (FVS), which counts each food item instead of food group, would be a more sensitive indicator for individuals with more or less diverse diets within a limited array of food groups (Rizvi 1991; Hatloy, Torheim, and Oshaug 1998; Finnis 2006). Food variety scores can, in certain contexts, provide a more in-depth analysis of dietary complexity, even when a limited number of food groups are being consumed. It may also be useful in illuminating intrahousehold variations in the range of foodstuffs being eaten within a given food group.

Dietary Transitions

Considerable attention is currently being paid to the health and economic consequences of the dietary transitions that accompany rural-to-urban migrations and changes in work patterns. In particular, there is increasing concern with changing modes of economic behavior and the regulation of daily activity patterns and energy expenditures. These changes, when accompanied by the spread of a "modern" diet, can contribute to rising rates of overnutrition and obesity in developing countries around the world, including Nepal. For example, Smith (1998, 1999) has demonstrated deleterious urbanization–food–activity level relationships among more affluent urban Nepalis who have been able to capitalize on the tourism-trekking trade.

As Popkin (2001) points out, overnutrition and undernutrition can coexist in one population, and is shaped in part by different livelihood contexts. By focusing on the trend towards globalizing dietary practices that feature "modern" food patterns and overnutrition/obesity, however, we risk overlooking how urbanization and changing activity patterns can contribute to chronic undernutrition, both in terms of the quality and the quantity of foods consumed.

This outcome has been demonstrated via the analysis of nutritional status among the children of the periurban carpet factory workers who participated in this research. When rural inhabitants move to marginal urban locations, they experience processes of dietary delocalization in that they are no longer able to produce their food. High-altitude, rural farming economies are no guarantee of nutritional adequacy. However, rural participants in this study were, at least seasonally, able to access more diverse foods because they raised livestock and planted traditional crops suited to the high-altitude environment. In contrast, while a variety of grains might be available in the urban marketplace, rice is the dominant grain in the diet; meat and animal products such as eggs and dairy are prohibitively expensive in an urban setting, particularly when migrants' earnings are low and tied to external, uncontrollable factors such as the peak carpet-buying season in foreign countries and the international tourist trade. Thus, while rural-to-urban migration means exposure to a greater range of foods than that available in rural areas inaccessible by road, it does not mean that these foods are more readily accessible in a practical, financial sense. The majority of the children living in the carpet-factory setting have low dietary diversity scores, and it appears that, given the expense of meat and dairy products, money is spent on cheaper, less nutritionally valuable items such as commercial bread, noodles, and sweets and some more selective processed items such as Lactogen and Cerelac. Moreover, all of the women in the urban sample were working in the carpet-making industry, and did not have the time to engage in making labor-intensive foods that are produced in the villages. They relied, when possible, on preprocessed, store-purchased products. Thus, urban children miss out on specific traditional foods that are tied to a rural farming economy, including home-grown barley porridge and salt butter tea.

The research on dietary transition and commoditization demonstrates that the impact on nutritional status varies from one case to another. A review of food commoditization by Dewey (1989) in Peru, Jamaica, and Mexico concluded that dietary diversity, quality, and nutritional status is usually better if home-produced foods are included in the diet. Similarly, Leonard et al. (1993), in a comparison between highland subsistence and coastal cash-cropping farmers in Ecuador, reported lower-quality diets among the latter group. Conversely, dietary studies in the Peruvian Andes found that, in general, dietary diversity and caloric intake increased in semiurban areas relative to rural communities due to access to market foods; but with the caveat that dietary diversity was much lower among low SES than the middle SES families who were better able to afford market foods (Leonard 1992; Leatherman 1994). Thus, it is clear that economic transition does not lead to one inevitable dietary outcome. As Leonard (1992) and Leatherman (1994) point out, local variation must be taken into account, as dietary transitions do not homogenously impact a population. Moreover, the previous studies cited focused on household or adult-only samples. More research on how dietary transitions differentially affect children's and adults' diets should also be considered, as outcomes may differ within households.

Moral Economies of Food

At this point, we briefly consider another way that dietary diversity analyses can provide insights into the biocultural aspects of human diet and nutrition. Dietary transitions are more than simple shifts in nutritional intakes. Food consumption practices also inevitably encompass values, preferences, and local experiences. By focusing solely on *what* people are eating, we risk overlooking *how* people feel about the food they eat and what values are placed on changing food consumption patterns. That is, dietary transitions also encompass the experiential aspects of food that exist in local contexts and at the individual level. This latter type of analysis provides a potential emerging research agenda that could be explored in order to better understand the ways that dietary transitions are experienced on a day-to-day basis.

In undertaking an analysis of this data, we began to ask questions that the dietary recall data were not equipped to answer. These were questions about preferences and the values placed on different kinds of foods and different dietary patterns. In the context of a process of urbanization, do mothers consider rural foods healthier for their children? How might these perceptions intersect with quantitative analyses of children's nutritional and health statuses? Are prepackaged breads and noodles given high social value, or are they necessarily part of an urban lifestyle where wage labor outside of the home is the case for many women? We also began to wonder how food values and preferences might reflect broader dietary transitions, affect food acquisition practices, and influence the foods that children are fed.

In order to understand the social impacts of dietary transitions, it is thus important to consider issues of food preference, beliefs about the health aspects of various kinds of food, the social meanings of foods, food cravings and longings, and ease of preparation issues. These experiential factors are part of the highly localized moral economies of food that can be linked to economic context, personal histories, and food symbolism. While there is considerable research examining various symbolic aspects of food in both contemporary and historical settings (e.g., Douglas 1972; Nichter and Nichter 1983; Mintz 1985; Khare and Rao 1986; Weismantel 1989; Cook and Crang 1996; Feldman 2005), purely quantitative assessments of nutritional status and dietary patterns may risk overlooking food symbolism, preferences, beliefs, and feeding practices in favor of prioritizing numerical, measurable results (but see Gittlesohn 1991; Moffat 2002; Finnis 2006). Yet, how people feel about the dietary transitions they are experiencing and the specific types of foods they are consuming has enormous potential implications for understanding such shifts and addressing policy issues (Shah 1983). That is, in analyzing the quantitative nutritional data gathered in the field, it became clear that it is necessary to better understand the specific contexts within which people in Nepal and elsewhere experience dietary transitions, and the ways that these changes do or do not reflect economic practices and decisions, as well as folk notions of nutrition and health. For example, have high-altitude migrants abandoned eating barley in favor of rice for reasons of prestige and/or accessibility? This is about more than context alone; it also includes giving more weight to experiential, preferential aspects of food practices in an analysis of diet and nutrition (see Casiday et al., this volume, for an example of this type of research).

Eliciting such qualitative understandings entails using methodologies and approaches that privilege local voices, providing for a more holistic and complex analysis of local experiences of dietary transitions. This includes considerations of food preferences and perceptions of food-health relationships, the status and sense of satisfaction associated with consuming different kinds of food, and the contexts that shape access to food. One way to approach these kinds of questions involves investigating the shared memory of food, those local histories that are associated with food and food changes. Such an approach engages with notions of agency and food choice, recognizing that dietary transitions may not simply be external processes shaped by international food practices, economic circumstances, migration, livelihood changes, and labor issues. There may also be elements of food choice that elude quantitative analysis techniques. As Shah (1983) has pointed out, expensive, less-nutritious foods may be preferred among low-income families as a result of food trends among high-income households (see also Weismantel 1989). This is particularly important in research that is policy oriented; otherwise, local voices and context may be lost under the weight of discussions that rely on Western nutritional indices, quantitative methods of analysis, and the discussion of broad, national, and regional-scale dietary shifts.

Conclusion

Though there are still some methodological issues that require consideration, dietary diversity as measured by the Dietary Diversity Score (DDS) is a useful and relatively simple means of investigating dietary quality and transition. It is crucial to consider dietary transitions in local context. While most of the current concern in developing countries is the transition to high-fat, high-sugar, and low-fiber diets, we must continue to consider the persistence of undernutrition, particularly for children. As illustrated by this periurban case study, dietary transitions from small-scale farming to urban wage-labor economies with concomitant delocalization of food and an increase in preprocessed foods can result in low dietary diversity and low intake of animal food products. In order to assist in improving the diets of rural-to-urban migrant children, however, we need to understand more than just what children are being fed by their families. We also must consider how and why caregivers make the choices they do. This requires research into food acquisition, preparation, and consumption practices that explore peoples' values, preferences, and everyday circumstances.

Notes

1. Growth stunting is defined as height that is less than two standard deviations below the median height for children of the same age and sex in a reference population (Gibson 1990), in this case the National Centre for Health Statistics (NCHS) survey from the United States (Hamill et al. 1979).
2. This is not necessarily the case in countries such as the United States, where a limited diet may center on cheap, processed foods that are high in calories and fat (see Randall, Nichaman, and Contant, Jr. 1985).
3. It is possible that urban children also consumed a food in the oil and fat category, perhaps in generous amounts of oil or ghee used for cooking; however, since cooking methods were not investigated in the food frequency recall, we were not able to include that food group in the DDS for the urban sample.

References

A. D. B. 2000. *Nepal Urban Sector Strategy*, vol. 1, *Main text*. TA: 3272. Kathmandu: His Majesty's Government of Nepal and the Asian Development Bank.

Allen, L. H. 1994. Nutritional influences on linear growth: A general review. *European Journal of Clinical Nutrition* 48: S75–S89.

Arimond, M., and M. T. Ruel. 2004. Dietary diversity is associated with child nutritional status: Evidence from 11 demographic health surveys. *The Journal of Nutrition* 134: 2579–85.

Cachel, S. 1997. Dietary shifts and the European Upper Palaeolithic transition. *Current Anthropology* 38 (4): 579–603.

Cook, I., and P. Crang. 1996. The world on a plate: Culinary culture, displacement and geographical knowledges. *Journal of Material Culture* 1: 131–53.

Costello, A. M. 1989. Growth velocity and stunting in rural Nepal. *Archives of Disease in Childhood* 64: 1478–82.

Dewey, K. G. 1981. Nutritional consequences of the transformation from subsistence to commercial agriculture in Tabasco, Mexico. *Human Ecology* 9 (2): 151–87.

———. 1989. Nutrition and the commoditization of food systems in Latin America and the Caribbean. *Social Science and Medicine* 28 (5): 415–24.

Diaz, J. R., A. de las Cagigas, and R. Rodriguez. 2003. Micronutrient deficiencies in developing and affluent countries. *European Journal of Clinical Nutrition* 57: S70–S72.

Douglas, M. 1972. Deciphering a Meal. *Daedalus* 101 (1): 61–81.

Feldman, C. 2005. Roman Taste. *Food, Culture and Society* 8: 7–31.

Finnis, E. 2006. "The Political Ecology of Crop Commercialization and Dietary Change in the Kolli Hills, South India." PhD dissertation. Hamilton, ON, McMaster University.

———. 2007. The political ecology of dietary transitions: Agricultural change, environment, and economics in the Kolli Hills, India. *Agriculture and Human Values* 24 (3): 343–53.

Fleuret, P., and A. Fleuret. 1980. Nutritional implications of staple food crop successions in Usambara, Tanzania. *Human Ecology* 8 (4): 311–27.

Foote, J. A., S. P. Murphy, L. R. Wilkens, P. P. Basiotis et al. 2004. Dietary variety increases the probability of nutrient adequacy among adults. *Journal of Nutrition* 134: 1779–85.

Fricke, T. E. 1993. *Himalayan Households: Tamang Demography and Domestic Processes.* Delhi: Book Faith India.

Gibson, R. S. 1990. *Principles of Nutritional Assessment.* New York: Oxford University Press.

Gittelsohn, J. 1991. Opening the box: Intrahousehold food allocation in rural Nepal. *Social Science and Medicine* 33: 1141–54.

Haddad, L. 2003. Redirecting the diet transition: What can food policy do? *Development Policy Review* 21 (5–6): 599–614.

Hamill, P. V. V, T. A. Drizd, C. L. Johnson, R. B. Reed et al. 1979. Physical growth: National Centre for Health Statistics percentiles. *The American Journal of Clinical Nutrition* 32: 607–29.

Harpham, T., T. Lusty, and P. Vaughn. 1990. *In the Shadow of the City: Community Health and the Urban Poor.* Oxford: Oxford University Press.

Hatloy, A., L. E. Torheim, and A. Oshaug. 1998. Food variety—a good indicator of nutritional adequacy of the diet? A case study from an urban area in Mali, West Africa. *European Journal of Clinical Nutrition* 52 (12): 891–98.

Hijazi, N., A. Bahaa, and A. Seaton. 2000. Diet and childhood asthma in a society in transition: A study of urban and rural Saudi Arabia. *Thorax* 55: 775–79.

Hoddinott, J., and Y. Yohannes. 2002. *Dietary Diversity as a Household Food Security Indicator.* Washington, DC: Food and Nutritional Technical Assistance Project, Academy for Educational Development.

Huijbers, P. M. J. F., J. L. M. Hendriks, W. J. M. Gerver, P. J. De Jong et al. 1996. Nutritional status and mortality of highland children in Nepal: Impact of sociocultural factors. *American Journal of Physical Anthropology* 101: 137–44.

Kedia, S. 2004. Changing food production strategies among Garhwali resettlers in the Himalayas. *Ecology of Food and Nutrition* 43: 421–42.

Kennedy, G. L., M. R. Pedro, C. Seghieri, G. Nantel et al. 2007. Dietary diversity score is a useful indicator of micronutrient intake in non-breast-feeding Filipino children. *The Journal of Nutrition* 137: 472–77.

Khare, R. S., and M. S. A Rao, eds. 1986. *Food, Society and Culture: Aspects of South Indian Food Systems.* Durham, NC: Carolina Academic Press.

Leatherman, T. L. 1994. Health implications of changing agrarian economies in the Southern Andes. *Human Organization* 53 (4): 371–80.

Leonard, W. R. 1992. Variability in adaptive responses to dietary change among Andean farmers. In *Health and Lifestyle Change,* ed. R. Huss-Ashmore, J. Schall, and M. Hediger, 71–82. MASCA Research Papers, vol 9.

———. 1994. Evolutionary perspectives on human nutrition: The influence of brain and body size on diet and metabolism. *American Journal of Human Biology* 6 (1): 77–88.

Leonard, W. R., K. M. DeWalt, J. E. Uquillas, and B. R. DeWalt. 1993. Ecological correlations of dietary consumption and nutritional status in highland and coastal Ecuador. *Ecology of Food and Nutrition* 31: 67–85.

Lunn, P. G. 2000. The impact of infection and nutrition on gut function and growth in childhood. *Proceedings of the Nutrition Society* 59: 147–54.

Martorell, R., and T. J. Ho. 1984. Malnutrition, morbidity and mortality. *Population and Development Review* 10 (Supplement): 49–68.

Maunder, E. M. W., J. Matji, and T. Hlatshwayo-Molea. 2001. Enjoy a variety of foods—difficult but necessary in developing countries. *South African Journal of Clinical Nutrition (Supplement)* 14 (3): S7–S11.

Messer, E. 1984. Anthropological perspectives on diet. *Annual Review of Anthropology* 13: 205–49.

Mintz, S. 1985. *Sweetness and Power: The Place of Sugar in Modern History.* Harmondsworth, UK: Penguin.

Moffat, T. 2001. A biocultural investigation of the weanling's dilemma in Kathmandu, Nepal: Do universal recommendations for weaning practices make sense? *Journal of Biosocial Science* 33: 321–38.

———. 2002. Breastfeeding, wage labor, and insufficient milk in peri-urban Kathmandu, Nepal. *Medical Anthropology* 21: 165–88.

———. 2003. Diarrhea, respiratory infections, protozoan gastrointestinal parasites and child growth in Kathmandu, Nepal. *American Journal of Physical Anthropology* 122 (1): 85–97.

Molleson, T., K. Jones, and S. Jones. 1993. Dietary changes and the effects of food preparation on microwear patterns in the Late Neolithic of abu Hureyra, northern Syria. *Journal of Human Evolution* 24: 455–68.

Naborro, D., P. Howard, C. Cassels, M. Pant et al. 1988. The importance of infections and environmental factors as possible determinants of growth retardation in children. In *Linear Growth Retardation in Less Developed Countries. Nestle Nutrition Workshop Series* vol. 14., ed. J. C. Waterlow, 165–84. New York: Raven Press.

Nichter, M., and M. Nichter. 1983. The ethnophysiology and folk dietetics of pregnancy: A case study from South India. *Human Organization* 42 (3): 235–46.

O'Neill, T. 2001. "Selling girls in Kuwait": Domestic labor migration and trafficking discourse in Nepal. *Anthropologica* 43: 153–64.

Panter-Brick, C. 1997. Seasonal growth patterns in rural Nepali children. *Annals of Human Biology* 24 (1): 1–18.

Panter-Brick, C., A. Todd, and R. Baker. 1996. Growth status of homeless Nepali boys: Do they differ from rural and urban controls? *Social Science and Medicine* 43 (4): 441–51.

Panter-Brick, C., P. G. Lunn, R. M. Langford, M. Maharjan et al. 2008. Pathways leading to early growth faltering. *British Journal of Nutrition* 101: 550–59.

Pelto, G .H., and P. J. Pelto. 1983. Diet and delocalization: Dietary changes since 1750. *Journal of Interdisciplinary History* 14 (2): 507–28.

Popkin, B. M. 1993. Nutritional patterns and transitions. *Population and Development Review* 19 (1): 138–57.

———. 1994. The nutrition transition in low-income countries: An emerging crisis. *Nutrition Reviews* 52 (9): 285–98.

———. 2001. The nutrition transition and obesity in the developing world. *Journal of Nutrition* 131: 871S–873S.

Popkin, B. M., S. Horton, S. Kim, M. Mahal et al. 2001. Trends in diet, nutritional status, and diet-related noncommunicable diseases in China and India: The economic costs of the nutrition transition. *Nutrition Reviews* 59 (12): 379–90.

Randall, E., M. Z. Nichaman, and C. F. Contant, Jr. 1985. Diet diversity and nutrient intake. *Journal of the American Dietetic Association* 85: 830–36.

Raphael, D., and H. Delisle. 2005. Households with undernourished children and overweight mothers: Is this a concern for Haiti? *Ecology of Food and Nutrition* 44: 147–65.

Rizvi, N. 1991. Socioeconomic and cultural factors affecting interhousehold and intrahousehold food distribution in rural and urban Bangladesh. In *Diet and Domestic Life in Society*, ed. A. Sharman, J. Theophano, K. Curtis, and E. Messer, 91–118. Philadelphia: Temple University Press.

Ruel, M. T. 2003. Operationalizing dietary diversity: A review of measurement issues and research priorities. *Journal of Nutrition* 133: 3911S–3926S.

Savy, M., Y. Martin-Prevel, P. Sawadogo, Y. Kameli et al. 2005. Use of variety/diversity scores for diet quality measurement: Relation with nutritional status of women in a rural area in Burkina Faso. *European Journal of Clinical Nutrition* 59: 703–16.

Schmidt, C. W. 2001. Dental microwear evidence for a dietary shift between two nonmaize-reliant prehistoric human populations from Indiana. *American Journal of Physical Anthropology* 114: 139–45.

Shack, K. W., L. E. Grivetti, and K. G. Dewey. 1990. Cash cropping, subsistence agriculture, and nutritional status among mothers and children in lowland Papua New Guinea. *Social Science and Medicine* 31 (1): 61–68.

Shah, C. H. 1983. Food preference, poverty and the nutrition gap. *Economic Development and Cultural Change* 31: 121–48.

Smil, V. 2002. Eating meat: Evolution, patterns, and consequences. *Population and Development Review* 28 (4): 599–639.

Smith, C. 1998. Prevalence of obesity and contributing factors among Sherpa women in urban and rural Nepal. *American Journal of Human Biology* 10: 519–28.

———. 1999. Blood pressures of Sherpa men in modernizing Nepal. *American Journal of Human Biology* 11: 469–79.

Smith, P. A., and R. M. Smith. 1999. Diets in transition: Hunter-gatherer to station diet and station diet to the self-select store diet. *Human Ecology* 27 (1): 115–33.

Steyn, N. P., J. H. Nel, G. Nantel, G. Kennedy et al. 2006. Food variety and dietary diversity scores in children: Are they good indicators of dietary adequacy? *Public Health Nutrition* 9 (5): 644–50.

Torheim, L. E., I. Barikmo, A. Hatloy, M. Daikite et al. 2001. Validation of a quantitative food-frequency questionnaire for use in Western Mali. *Public Health Nutrition* 4 (6): 1267–77.

Weismantel, M. 1988. *Food, Gender, and Poverty in the Ecuadorian Andes*. Philadelphia: University of Pennsylvania Press.

———. 1989. The children cry for bread: Hegemony and the transformation of consumption. In *The Social Economy of Consumption*, ed. B. Orlove and H. J. Rutz. Lanham, MD: University Press of America.

• 7 •

Responses to a Food Crisis and Child Malnutrition in the Nigerien Sahel

R. E. Casiday, K. R. Hampshire, C. Panter-Brick, and K. Kilpatrick

Introduction

Food insecurity, poverty, and risks to health are clearly political and economic issues requiring top-level intervention. An important question is whether intervention efforts launched by humanitarian or governmental structures articulate effectively with the grassroots strategies developed by local communities to face acute, chronic, or acute-upon-chronic food shortages. In addressing food crises and/or chronic poverty, humanitarian agencies, central governments, local communities, and individual households may work towards the same critical aims—to protect lives and livelihoods—yet operate within different sets of constraints, worldviews, and expectations. For instance, humanitarian agencies have a mandate to step up all efforts in order to manage a transitory "emergency" situation, such as a food crisis on an unprecedented scale, because such crisis demands immediate action to "save lives." This level of intervention contrasts with longer-term efforts for managing the "silent emergency" of child malnutrition in communities that suffer chronic, albeit severe, poverty. Local households, however, often cope with severe food shortages as part and parcel of habitual experience; their responses to an acute food crisis are not necessarily *qualitatively* different from the coping responses to everyday poverty, reflecting their dual concern to save both fragile lives and vulnerable livelihoods.

In this chapter, we draw upon the example from the Nigerien Sahel at a time of severe food shortage to review macro- and micro-level responses to food insecurity, clarifying the priorities that guide them. We examine how risks to health and survival are operationalized by humanitarian agencies involved in sub-Saharan famine relief, with reliance on biological (anthropometric and mortality) indicators of child vulnerability to differentiate between different levels of emergency requiring intervention. We also document local responses vis-à-vis emergency nutrition and health programs, how these are embedded in social experience, and what they reveal about indigenous frameworks for understanding vulnerability and resilience in the context of chronic poverty.

Food Insecurity and Grinding Poverty in the Poorest Country of the World

Niger is a landlocked country in the Sahel region of Central West Africa, currently ranked as the poorest country in the world on the UN Human Development Index (UNDP 2006). Out of a population estimated at 12.4 million in 2005, 63 percent live on less than a dollar a day—and 34 percent live in extreme poverty (Mousseau and Mittal 2006).

Recent decades have seen very high population growth (3.3 percent annually between 1988 and 2001) and increasing overexploitation of land for agricultural production (World Bank 2002; May, Harouna, and Guengant 2004). Population pressure, land degradation, reduced income-generating opportunities, and weak markets all conspire to render the people of Niger vulnerable to food insecurity (Baro and Deubel 2006). Poor harvests result in price explosions of basic grains, making them unaffordable to the poor. Many are forced to borrow and repay at much higher rates after the harvest, creating cycles of indebtedness and food insecurity. Moreover, poor access to health care, high rates of infection and illness, and inappropriate infant feeding practices contribute to worsen the effects of food scarcity on young children (Daulaire 2005; Hampshire, Casiday et al. 2009).

In 2005, recurrent food insecurity in the Sahel caught international attention, as its impact on human lives and livelihoods reached a crisis point—warranting "emergency intervention" (Figure 7.1). A severe drought and locust invasion had led to very poor harvests and widespread food shortages in 2004. Although total food production for Niger was only 7.5 percent below the national food requirement, control of the national cereal markets by a few big traders meant that many people were unable to afford the high food prices that year (Mousseau and Mittal 2006). The situation failed to improve in the following year, because communities were left in a precarious position, their livelihoods at risk from having sold their livestock and incurred debts to cope with the food crisis of the preceding year (Daulaire 2005; Kapp 2005).

In some localities, however, the "food crisis" of 2004–2006 was not necessarily qualitatively different from other years. Seasonal hunger and food crises are part and parcel of habitual, "normal" experience. It is even argued that it is the very nature of

Figure 7.1. Affala, Niger, October 2005. Families participating in the Concern Worldwide emergency nutrition program queue to receive their 'family ration' of grain, pulses, and oil. Photograph courtesy of Kate Kilpatrick.

chronic hunger in this part of the world that resulted in delayed and inadequate responses, on the part of the Nigerien government and of international donors, to the emergency situation of 2004–05 (Mousseau and Mittal 2006).

Appraisal of a Food Crisis and Child Malnutrition

For Western outsiders, Niger was clearly in an emergency situation during the 2005 Sahelian food crisis. Many Nigerien people endured extreme conditions—during our interviews in 2006; for example, one woman recounted being prostrate on the ground, too weak to get up to meet the Western photographers who were "clicking and flashing at my face." Another woman, whom we asked how villagers had been able to complete the fasting month of Ramadan, replied, "Ramadan or not, it made no difference—we had nothing to eat."

Despite the sudden media attention, some critical reviewers have emphasized that this was "not a transitory emergency but a permanent feature of mounting vulnerability" (Baro and Deubel 2006: 529). Indeed Rubin (2006) argued that the country was in fact facing two emergencies—one immediate and the other long term—and that both demanded intervention.

Humanitarian agencies rely on a standardized set of indicators to classify food crisis situations requiring different levels of humanitarian assistance (SPHERE 2004; Young and Jaspars 2006). Data collection surveys and intervention efforts frequently focus on young children, six months to five years of age. Children under five have high nutritional requirements per kg body weight and require nutritionally dense foods. These children are also vulnerable to infectious diseases, especially during the weaning process and when they begin to crawl on dirty ground. Nutritional supplementation of babies under six months of age presents greater difficulties, given global frameworks of breastfeeding promotion that recommend exclusive breastfeeding for infants under six months (WHO 2001).

The most common indicators used to assess the severity of food crises are community prevalence of acute malnutrition, together with child mortality rates and other aggravating factors such as infectious disease epidemics (Table 7.1). Acute malnutrition indices rely on measures of child growth expressed as z-scores (standard deviations, SD) from the median of a reference population for children of the same sex and age. Global acute malnutrition (GAM) is taken as the percentage of children with weight-for-height z-scores (WHZ) ≤ -2 relative to the reference population; severe acute malnutrition (SAM) is taken as WHZ ≤ -3. Different agencies have developed their own frameworks, which differ slightly in terms of definitions and thresholds (Table 7.1). According to the WHO framework, special interventions are required where GAM exceeds 15 percent (indicating a critical situation for community rates of child malnutrition) or 10 percent (indicating a serious situation), while the presence of aggravating factors serves to trigger interventions from lower thresholds of rates of acute malnutrition (Table 7.1). It should be noted that these frameworks offer guidelines only, and may need to be adapted for local situations (Young and Jaspars 2006: 23).

Mid–upper-arm circumference (MUAC) is another helpful indicator of child nutritional status. It changes little between one and five years of age, has a stronger association with mortality risk than weight-for-height within this age group (Young and Jaspars 2006), and is often used as a criterion for admission to therapeutic feeding programs for acutely malnourished children (with MUAC < 110 mm). While MUAC is often reported in nutrition surveys alongside WHZ, the two indicators may identify different children as malnourished; caution should be employed when comparing prevalence rates of malnutrition based on one indicator against surveys using a different indicator.

In Niger, child health and mortality rates are certainly very problematic (Table 7.2), reflecting both a crisis situation and the underlying perennial problem of grinding poverty. Child malnutrition rates regularly fall in the risky to serious categories of the WHO framework (Table 7.1), representing "a chronic nutritional emergency" (Mousseau and Mittal 2006: 3). Nutritional surveys conducted in early 2005 found GAM rates in excess of 20 percent, far above the threshold defining an emergency situation. In addition to acute malnutrition (GAM and SAM) in 2004–05, there is plentiful

Table 7.1. International frameworks and indicators of a food crisis.

	Médecins Sans Frontières framework (MSF 1995)*,	World Health Organization framework (WHO 2000)*,	Action required
Critical/ serious situation	>20% GAM among children 6 months to 5 years -OR- 10–19% GAM with aggravating factors[1]	>15% GAM among children 6 months to 5 years -OR- 10–14% GAM with aggravating factors[1]	• General rations • Supplementary feeding for all members of vulnerable groups • Therapeutic feeding for severely malnourished individuals
Risky situation	10–19% GAM -OR- 5–9% GAM plus aggravating factors[1]	10–14% GAM -OR- 5–9% GAM plus aggravating factors[1]	• Supplementary feeding for malnourished individuals in vulnerable groups • Therapeutic feeding for severely malnourished individuals
Acceptable situation	<10% GAM without aggravating factors[1]	<10% GAM without aggravating factors[1]	• Attention to malnourished individuals through regular community services

*Adapted from Young and Jaspars (2006: 23). GAM = global acute malnutrition. WHZ (Weight-for-height z-score) ≤ -2.
[1] e.g. excessive child mortality, epidemic of communicable diseases

evidence of chronic malnutrition (as indicated by stunting, or height-for-age z-scores). Child indicators of malnutrition clearly reveal an acute-upon-chronic situation.

Government, Donor, and Humanitarian Responses

The situation in Niger soon raised a very sensitive question, best phrased by Mousseau and Mittal (2006: 14) in these terms: "Why was the poorest country in the

Table 7.2. Child health and nutrition indicators for Nigerien children (6–59 months): malnutrition (growth status), infections (preceding two weeks), and mortality.

	Niger, Surveys from 1996 to 2005[1]	Niger, 1998 DHS[2]	Niger, 2006 DHS[3]	Tahoua and Illéla Districts, Niger, Survey in Dec 2005[4]	Tahoua and Illéla Districts, Niger, Survey in Dec 2006[5]
Malnutrition rates (%)					
Wasting: Global Acute Malnutrition (GAM): Weight-for-height z-scores ≤ −2	13.4–24.0	20.7	10.3	14.8	Tahoua: 8.1 Illéla: 9.5
Wasting: Severe Acute Malnutrition (SAM), Weight-for-height z-scores ≤ −3	1.6–5.4	3.7	1.5	2.4	
Stunting: Global Chronic Malnutrition (GCM), Height-for-age z-scores ≤ −2		38.0	50.0	38.0	
Infection rates (%)					
Fever		48.3		66	
Diarrhea		37.8		11	
Acute Respiratory Infections		14.2		10	
Mortality					
Daily deaths (number deaths per 10,000 before age 5)				4.3	2.1
$_5q_0$ (risk of dying before age 5, per 1,000)	257				

Sources: [1]FEWS NET, 2006; [2]DHS 1998; [3]INS and Macro International Inc., 2007; [4]Kokere 2006; [5]Concern Worldwide 2007.

world, with one of the highest levels of malnutrition, so unsuccessful in attracting donors' interest in 2005?" These authors argue that responsibility for this "lies first with the government, which failed to request an adequate level of assistance," but that this crisis had also been "largely ignored by donors" (Mousseau and Mittal 2006: 14).

The Nigerien government had first responded with a very limited call for international assistance. Its system of prevention and response to food crises (consisting largely of maintaining a national grain reserve) had been in place since the 1980s; yet in 2004, the government had only 20,000 tons of grain at its disposal, and financial reserves to procure an additional 20,000 tons—against a national deficit of 505,000 tons (Mousseau and Mittal 2006)! In November 2004, the government requested 32.7 million euros (US $42 million) in additional support, to be used primarily for the procurement of 78,000 tons of food to be sold to those in need at subsidized prices. Not only was this amount woefully inadequate to address the scale of the problem, but less than 10 percent of requested funds had actually been donated by July 2005. Moreover, Médecins Sans Frontières denounced the strategy of offering only subsidized (rather than free) food to the poor, because many could not afford even the subsidized prices and were continuing to starve (Jezequel and Yzebe 2005). Food-for-work and food-for-loan schemes were later implemented, although these also had significant problems: in particular, borrowing food meant incurring debt when grain prices were at their highest and repaying it at four to five times the borrowed amount after the harvest, when prices were lowest (Eilerts 2006).

In their detailed case study of the Nigerien food crisis, Mousseau and Mittal stated that donors were finally stirred to action, after news of the famine had been broadcast to the world. "In the face of this mounting crisis, international NGOs raced to set up a huge intervention in order to treat 230,000 malnourished children under five—an astonishing number, far higher than any previous relief intervention" (Mousseau and Mittal 2006: 13). Yet a comparison of international assistance shows how few funds were devoted to Niger, relative to other emergencies: for example, Niger received twenty-five times less aid from the United States than did Darfur ($6 vs. $149 per person affected, respectively), even though the absolute numbers of people affected were much the same (3.2 million in Niger, 3.4 million in Darfur; Mousseau and Mittal, 2006: 14).

The reasons for the limited governmental and international donor responses are complex and, to some extent, obscure. Niger's long-term poverty and reliance on foreign aid (with half the national budget dependent on international donors) mean that policy decisions are largely dictated by the development strategies of the World Bank and other funders, requiring recipient countries to implement structural adjustment programs to encourage private-sector economic growth (Jezequel and Yzebe 2005; Drouhin and Defourny 2006; Mousseau and Mittal 2006). Free food distribution, it was argued, would disrupt local and national markets (undercutting farmers' livelihoods) and thus hamper the country's long-term economic development.

The government was fighting significant resourcing issues, limiting its ability to respond to a scaled-up version of donor operations, but was also concerned with preserving a sense of national dignity. In September 2005, Prime Minister Hama Amadou complained: "Our dignity suffered. And we've seen how people exploit images to pledge aid that never arrives to those who really need it" (BBC News Online 2005). The government became extremely sensitive to the charge that it had allowed sections of the population to become critically malnourished. Low expectations may also have played a role in the inadequate response: the limited request for aid perhaps reflected the Nigerien government's (realistic) assessment of what international donors were actually willing to pay for. By contrast, the military coup that overthrew the government in 2010 ushered a new transparency: one of the first policy decisions was to appeal for $123 million in foreign aid to tackle the still-existing food crisis (Tsai, 2010).

On the part of donors, delay may also have been fueled by a lack of confidence in the Nigerien government and shortage of existing, on-the-ground relief agencies, which meant that Niger was not sufficiently visible on donor radar screens. More critical voices have argued that the chronic nature of hunger in Niger had made the situation appear inevitable, even "normal" (Drouhin and Defourny 2006, Mousseau and Mittal 2006: 4).

Several humanitarian agencies implemented programs responding in various ways to the crisis. We are most familiar with the work of Concern Worldwide, having had access to a number of consultancy reports. Concern implemented an emergency nutrition program, adopting a strategy of community-based therapeutic care (CTC) broadly in line with the Valid International protocols for CTC (Valid International 2006). Community workers based in villages were trained in the targeted referral of children. Local centers offered "supplementary feeding programmes" (SFP) for the treatment of moderate malnutrition, "outpatient therapeutic programmes" (OTP) for the treatment of severe malnutrition, and "onwards referral" to inpatient care for the treatment of severe malnutrition with medical complications in a regional stabilization center (SC).

In villages, children between six months and five years of age were screened for acute malnutrition using MUAC (mid–upper arm circumference). Children with MUACs of less than 110 mm were admitted directly on to the OTP or to the regional SC. Those with MUACs between 110 mm and 125 mm were referred to feeding centers, where height and weight were measured (Concern Worldwide 2007). Children presenting with moderate or severe malnutrition were admitted onto the SFP or OTP programs respectively. All CTC children were given supplementary rations of Unimix (enriched porridge) and vegetable oil, while severely malnourished children were given in addition between two and five daily rations of "Plumpy'nut" sachets (each 500 kcal), depending on their body weight. Plumpy'nut (Figure 7.2) is a specially formulated peanut butter–like product that is nutritionally similar to therapeutic milk formula and can be eaten directly from the package (thus eliminating concerns

Figure 7.2. Plumpy'nut, a peanut-based therapeutic food used for treating infants and children with severe acute malnutrition, is provided ready-for-use in small sachets. Photograph courtesy of Kate Kilpatrick.

about mixing the product with contaminated water), enabling outpatient treatment. In addition, all CTC children were given a "protection ration" of premixed corn-soy blend, oil, and sugar, and their families received a one-off "family ration" of grain, pulses, and oil, to minimize the sharing of high-density food supplements with better-nourished siblings. All children admitted to the emergency nutrition programs were also given basic medical treatment and referred to health centers for vaccination, as appropriate. Limited water and sanitation programs were also provided.

While many of the agencies involved were also running longer-term development programs, the emergency feeding programs were explicitly short-term relief efforts, targeted at saving the lives of the most vulnerable children as defined by anthropometric measures. As it has become increasingly clear that the problem of high levels of acute malnutrition in Niger is an ongoing one, there has been a shift in focus over recent months among humanitarian agencies towards integrating their programs with longer-term state-led health and development initiatives. For example, Concern Worldwide is now working closely with the Ministry of Health at national and regional levels to address some of the constraints that parents face in accessing curative healthcare for their children (Concern Worldwide 2007). While the government of Niger and humanitarian agencies both stress the importance of the integration of nutritional programming into wider health services, they face considerable operational and resource challenges in achieving sustainable improvements.

Experience on the Ground

How did intervention programs, deployed in those areas most affected by the food crisis and chronic poverty, articulate with the perspectives of local households? It is important to emphasize that Nigerien families were generally very enthusiastic and supportive of the emergency programs and that they recognized and appreciated the substantial benefits in terms of saving the lives of children who would otherwise have died. Indeed, one longer-term benefit of the programs might be to change parents' sense of agency and control over the fate of their children. In interviews, several parents commented on the fact that children who had appeared destined to die had recovered following timely and appropriate treatment.

Nonetheless, there were areas of disjuncture between local perspectives and the practices and experiences of the humanitarian programs. We focus on three aspects of local experience: (a) perspectives about vulnerability and entry into the emergency nutrition programs; (b) household allocation of food rations in relation to management of risk at individual and household level; and (c) leaving the program and concerns about sustainability.

We draw on data collected as part of an anthropological consultancy on behalf of Concern Worldwide in January–February 2006 (some three months after the harvest) in two rural districts of Niger: Tahoua and Illéla (Hampshire, Casiday et al. 2009; Hampshire, Panter-Brick et al. 2009). We used a range of qualitative methods, including participant observation, interviews, and consultation with government health workers or local staff employed by the humanitarian agencies. Sampling was purposive, including households with diverse child nutritional and health status (available from growth/health records), livelihood security, subsistence system, ethnic group, and distance from health services. Semistructured interviews were conducted with mothers (N=40), other caregivers (N=6), siblings (N=9), health workers and community leaders (N=38), in addition to focus groups with mothers and grandmothers (N=15 focus groups). We also collected morbidity and feeding histories, with seven-day dietary recall, for children under five from their primary caregivers (N=44), using structured interviews. Analysis of interview material was undertaken in consultation with local field assistants and key informants, enabling juxtaposition of emic and etic perspectives (Miles and Huberman 1994). This participatory approach to data analysis helps to ensure that data interpretations made sense locally (Young and Jaspars 1995b). In addition, we performed secondary analysis of nutritional and demographic survey data, collected by Concern (Kokere 2006).

Vulnerability and Entry into Feeding and Therapeutic Programs

While it was clearly necessary for humanitarian agencies to assess and target the most vulnerable children for assistance, noteworthy obstacles, including a disjuncture with some parental views of vulnerability, prevented some malnourished children from having access to emergency nutrition programs. Here, we give several examples to illustrate these problems.

The examples in Box 7.1 focus on babies less than six months old. Intervention efforts did not include young babies in supplementary feeding or outpatient therapeutic programs (SFP or OTP), although where appropriate, babies under six months could be admitted to the regional stabilization center (SC). Feeding young babies with supplementary foods presents complex problems, ranging from the suitability of liquid foods for an infant's biological requirement to the enhanced risk of illness with food supplementation, and the possibility of undermining good breastfeeding practice (World Health Organization 2001). The problem is that the prevalence of exclusive breastfeeding in Niger is extremely low, and that substantial numbers of babies with acute malnutrition are not included in programs of humanitarian intervention. Local surveys for Tahoua and Illéla Districts show that 0 percent and 11 percent of mothers, respectively, exclusively breastfed for the first four to six months of a baby's life (Kokere 2006). The most common reason for early cessation of breastfeeding is subsequent pregnancy. Weaning foods are introduced at a very young age (a matter of weeks), because some mothers believe that their breast milk is insufficient or "bad" on the grounds that babies become unhappy or ill. Thus breast milk is withdrawn or nutritionally poor supplementary foods are given just at a point when children are vulnerable to disease and malnutrition. Babies are also frequently given water-based preparations thought to have medical or health-giving properties.

Often, local people could not understand why acutely malnourished babies were not accepted at centers distributing supplementary food rations. We cannot estimate how many babies might have been affected, given that anthropometric data on children up to six months old are not available, but there was widespread perception of babies denied a right of entry. In the words of one Hausa mother: "I took him [four-month-old son] to the weighing, but they said he was too young. Look at him—he is so weak and I do not have enough milk to make him strong." This mother knew her baby was highly vulnerable, but the baby did not meet the age criteria for entry into the program.

A second issue concerns children who were not presented to feeding centers, even though they would be eligible for supplementary feeding, for a variety of reasons related to social status and ethnic identity. Consideration of social status is one reason why relative wealth does not appear to protect children from severe malnutrition (Young and Jaspars 1995a). As illustrated in case study C (Box 7.2), wealth and status can, under some circumstances, be obstacles to vulnerable children receiving assistance in the form of food supplements. Household expenditures to maintain the position of "big people" with considerable power over village-level decisions may at

Box 7.1.
Case studies of early/abrupt weaning of already vulnerable infants

Case A: Early supplementation due to perceived lack of breast milk

Zenabu (a Hausa mother) has seven living children and three dead children. Her youngest daughter, Hawa, was nine months old and malnourished at the time of interview. She began giving each of her children millet-water from the age of a few weeks, and here explains why:

Zenabu: All of my children were ill as young babies. They were born fine, but they all became ill when they drank my milk. I did not have enough milk in my breasts.

KH: How did you know you did not have enough milk?

Zenabu: I knew because my first two children died when they drank my milk. After they take my milk, they always cry and I know there is not enough for them. Look [points to Hawa]—see how small she is because I can't give her enough milk.

KH: So, when you found you didn't have enough milk, what did you do?

Zenabu: With the first two [children], I kept giving them my milk, and they both died. So, since then, I give them millet-water [the water that millet is cooked in, once it has been drained] from the time they are one week old.

Case B: Early cessation of breastfeeding due to perceived "bad milk"

Hadiza (a Hausa grandmother) explains why her granddaughter, Jamila, was weaned abruptly at the age of two months:

Hadiza: Jamila was born weak. Then, when she was two months old, she had difficulties breathing. I thought this had happened because her mother had fallen pregnant again, and so her milk had become bad.

KH: Did she say she was pregnant?

Hadiza: No, but many women do not want to say if they are pregnant. I thought her milk must be bad and that was why Jamila couldn't breathe properly. So I thought she must stop breastfeeding and I brought her back to live here with me.

first seem perverse, but could have far-reaching consequences for their well-being. Case D illustrates how powerful voices within the dominant social group were marginalizing other ethnic groups to harness exclusive access to humanitarian assistance and other development support. Finally, case E shows how relative wealth does little to protect children against risks of severe malnutrition, given detrimental childcare practices and the lack of available suitable weaning foods in the area.

Box 7.2
Considerations of wealth, status, and ethnic identity in childcare practices and access to emergency feeding programs

Case C: Malnourished child from a high-status family

Halima (a Tuareg mother) was from a very high-status family: she lived in a large courtyard and her husband was one of the big chiefs of the village, and yet her youngest daughter had been severely malnourished. It was clear that the family had been very wealthy, but had recently suffered a change of fortune. However, a strong sense of pride and shame constrained their responses to this newly found poverty.

Because he had never worked in his fields, Halima's husband continued to hire paid labor to cultivate, even when he could no longer afford to do so. He had to borrow money to pay the laborers, which he repaid by selling a substantial proportion of his grain—at very low postharvest prices. As a result, the household did not have enough millet to last through the year. When her daughter became very thin and ill, Halima refused to take her along to the weighing. She was too ashamed to admit that her child might be malnourished:

"How could I stand in the queue with all the other women from the village?"

Case D: Ethnic marginalization as a barrier to accessing food programs

Roukietou (a Mbororo mother) had a six-month-old son, Boureima, so emaciated he looked on the point of death. The Mbororo are a marginalized group of Fulani; this community lived a few kilometers away from a large

Hausa village. When asked if she had taken Boureima to the feeding center, Roukietou replied:

"No. When the program people come to [the village], the chief's wife chases us [Mbororo women] away."

Another woman corroborated:

"We always miss out on [NGO] programs, because we stay at the edge of the village."

A few minutes after this discussion, the chief's wife and other Hausa women came to chase the Mbororo women away, insisting the interviews should be done with Hausa women instead.

Case E: Severely malnourished twins in wealthy migrants

One of Bintou's twin children had been admitted for inpatient care at the regional stabilization center for treatment of severe malnutrition and medical complications. Bintou was dressed with good clothes and gold jewelery, a sign of substantial wealth. She had lived two years in Abijan, Côte d'Ivoire, where her husband had a business, but had now returned to live in the village for two years—it was her co-wife's turn to be taken by the head of household to Abidjan.

Bintou explained that because this child was born in the village, she had followed the local custom of giving *boule* (watery porridge) in the first few weeks of life and that did not make the baby strong. For one of her older children, born in Abidjan, she had followed the urban practice of breastfeeding for close to six months. She also remarked that in town, there were plentiful food supplements to give to young children, while in this area, there was nothing but *boule*.

The twin girl was discharged from the stabilization center, with rations of plumpy'nut to supplement her and rations of flour and oil for the family. Bintou said she would share the plumpy'nut between the twins—one acutely malnourished and recovering from malaria, the other malnourished but with a weight-for-height just under the threshold of admission qualifying her for individual emergency feeding rations.

Other practical factors, such as extremely long queues at the gates of feeding and therapeutic centers, prevented some people from accessing the emergency programs. This was particularly true for mothers who had busy work schedules or who were not articulate in Hausa. Mothers and grandmothers told us of having repeatedly tried to get children admitted to the local supplementary feeding center. One mother stated: "When I take him, there are so many people waiting—I never managed to get to the front of the queue, so I became disheartened." One Tuareg woman reported that she had been turned away after staff could not verify the age of her baby, because she could not express his date of birth in Hausa. These difficulties underscore the enormity of the problem faced by the humanitarian agencies in distributing emergency food rations, working with set criteria to manage an enormous number of people queuing for extra food rations.

Humanitarian relief agencies operate under immense fiscal, resource, and staffing constraints and so they must target their efforts to the most vulnerable, even if this means that others in need cannot always be helped. It is notoriously difficult in situations of food crisis to ensure that humanitarian aid is effectively delivered to the most in need. Poverty and malnutrition are commonly used as criteria for inclusion in emergency response programs but, particularly in complex emergencies, it has been argued that these do not always capture those most at risk (Jaspars and Shoham 1999).

Household Allocation of Rations and Management of Risk

The explicit goal of emergency feeding and therapeutic programs is to save the lives of the most vulnerable individuals. By contrast, parents seek to address the vulnerability of *all* of their children and the long-term livelihoods of the whole household (Hampshire, Panter-Brick et al. 2009).

There is a strong local ethos of nondiscrimination between children within a household. A widely expressed sentiment was that "all children are the same." This ethos is manifested in practices of food sharing and equal distribution of resources within households and applies to boys and girls equally, even when one child is sick or otherwise particularly vulnerable. Because there is no positive discrimination targeting food and health care to the most vulnerable children, the result is a de facto discrimination against them (Hampshire, Panter-Brick et al. 2009).

Sick children are not usually given special foods of high quality, or easy to digest. Indeed cultural practices of eating together and sharing food equally between children make it very difficult for parents to single out a child for special treatment. In addition, there is a strong belief that giving a child high quality foods, especially milk, for a temporary period can result in that child falling ill with a serious condition known as *anugu* (fever and rash) when the extra food is withdrawn. Fear of *anugu* underlies (or perhaps helps to justify) mothers' reluctance to give high-quality foods to sick children.

As we show elsewhere (Hampshire, Panter-Brick et al. 2009), it makes sense to spread risks in contexts where children's lives are seen as inherently precarious. There is a sense of powerlessness among parents over the fate of their children, such that even when extra resources are invested (high quality foods, health care, time), this will not necessarily guarantee the survival of that child. Furthermore, long-term livelihood insecurity means that families must maintain productive assets (such as livestock), as well as social capital, upon which the future well-being of the whole family depends.

In relation to the food supplements received from emergency feeding centers, parents' priorities do not, therefore, always coincide with those of the humanitarian agencies: parents do not always use rations for the purposes intended by donors. Despite the efforts of agencies to persuade parents against this, there were numerous accounts of parents dividing rations equally between all their children, rather than just giving them to the child identified as malnourished, as this focus group exchange between Hausa mothers illustrates:

"No one gives the ration just to the child it is meant for. We have to share between all children."

"And with the neighbors' children too, or we will be criticized for being selfish."

There were reports of people selling rations, as well as sharing them with others. One local schoolteacher offered us a sachet of Plumpy'nut that she had been given by the parents of one of her pupils. Such gifting behavior may seem perverse, but makes sense from a local perspective of risk-spreading to secure a range of social networks. Plumpy'nut is a highly desirable food. As such, respondents stated that it could be sold to pay for other food for the whole family. Plumpy'nut packets became currency as a way of acquiring social capital. By giving sachets to influential community figures, parents underscored important social networks that might later prove invaluable in times of crisis.

Leaving the Program: A "Second Weaning"

Finally, what happens to children who leave the feeding or therapeutic programs? Children are discharged when they have reestablished sufficient weight and health to leave at-risk categories. However, parents reported that their children had become accustomed to the food supplements and, in some cases, even refused to eat normal household foods (the bland, liquid porridge called *boule*) while they received the high density rations. Parents feared that, once the rations are withdrawn, children will be in a worse position than before. One woman compared the discharge from the feeding program to cessation of breastfeeding when a mother became pregnant, an abrupt weaning for the second time in the child's life. Another woman expressed similar concerns: "When Hawa [her baby] is better, they will stop the ration, and I am afraid that she will become ill again. That is what has happened to other babies here." This fear of potential illness, following sudden withdrawal of a food to which one has become accustomed, links into beliefs about *anugu*, as mentioned above.

Some children are readmitted to the supplementary feeding programs soon after their discharge, because they cannot thrive at home. Unfortunately, there are few quantitative or qualitative data on readmissions to emergency programs, an indicator that it has proven difficult to monitor accurately within such large-scale programs. Issues of discharge and readmission to emergency programs are areas of disjuncture between donor and local perspectives that remain little explored.

Conclusion

In tackling the problems of child undernutrition, sickness, and mortality in a food crisis situation, relief agencies, government, and households all share an overall goal—protecting children's lives while preserving livelihoods. However, they operate under very different constraints and thus adopt different strategies and priorities toward maximizing this goal. Humanitarian interventions in a "crisis situation" are hard to coordinate with government action: they achieve very substantial results in terms of saving individual lives, but frustrating results in terms of making those individual lives sustainable. Governments in very poor countries operate under formidable challenges to achieve equitable and efficient distribution of resources. Furthermore, local frameworks to spread risks and achieve resilience, in the face of chronic poverty and acute livelihood, food, and health crises, crucially impinge on the success of emergency and government interventions. Local and global constraints each make sense within their respective parameters, but there are points of disjuncture between them, especially in cases of acute-upon-chronic situations of mounting social and economic vulnerability.

To what extent are these different priorities reconcilable? In Niger, humanitarian agencies, such as Concern Worldwide, are increasingly working with government partners, such as the Ministry of Health, in addressing some of the barriers to treatment seeking for children's illnesses (particularly malaria, diarrhea, and acute respiratory infections) that can precipitate malnutrition, as well as working to train health workers in responding to malnutrition. Similarly, given the way that chronic livelihood insecurity underpins parents' risk-averse strategies of investment in vulnerable children, a top priority must be working to develop better linkages between humanitarian emergency programs and longer-term initiatives aimed at sustainable livelihood development (see Hampshire, Casiday et al. 2009). The issue of scaling-up nutrition programs to prevent malnutrition is debated at highest levels, e.g. at the World Bank (Blackwell, Augier et al. 2010). Particularly in cases of protracted crisis, it is essential to articulate effectively the position of humanitarian agencies, government structures, and local people, to achieve sustainable improvements in both child health and household livelihoods, and address the very real constraints of pervasive poverty.

References

Baro, M., and T. F. Deubel. 2006. Persistent hunger: Perspectives on vulnerability, famine, and food security in sub-Saharan Africa. *Annual Review of Anthropology* 35: 521–38.

BBC News Online. 2005. Niger food aid 'no longer needed.' http://news.bbc.co.uk/2/hi/africa/4253060.stm.

Blackwell, N., A. Augier and S. Sayadi. 2010. Food crisis in Niger: A chronic emergency. *The Lancet.* 376 (August 7): 416–417.

Concern Worldwide. 2007. *Emergency Nutrition Programme: Final Report to DEC ERP Tahoua and Illéla Districts 2006–07.* Concern Worldwide.

Daulaire, N. 2005. Niger: Not just another famine. *The Lancet* 366: 2004.

DHS. 1998. *Enquête Démographique et de Santé 1998.* CARE International and Macro International, Inc.

Drouhin, E., and I. Defourny. 2006. Niger: Taking political responsibility for malnutrition. *Humanitarian Exchange* 33: 20–22.

Eilerts, G. 2006. Niger 2005: Not a famine, but something much worse. *Humanitarian Exchange* 33: 17–19.

FEWS NET. 2006. *Understanding Nutrition Data and the Causes of Malnutrition in Niger: A Special Report by the Famine Early Warning Systems Network (FEWS NET).* FEWS NET.

Hampshire, K., R. Casiday, K. Kilpatrick, and C. Panter-Brick. 2009. The social context of childcare practices and child malnutrition during Niger's recent food crisis. *Disasters* 33 (1): 132–51.

Hampshire, K. R., C. Panter-Brick, K. Kilpatrick, and R. E. Casiday. 2009. Saving lives, preserving livelihoods: Understanding risk, decision-making and child health in a food crisis. *Social Science & Medicine* 68 (4): 758–65.

Institut National de la Statistique (INS) and Macro International Inc. 2007. *Enquête Démographique et de Santé et à Indicateurs Multiples du Niger 2006.* Calverton, Maryland, USA: INS and Macro International Inc.

Jaspars, S., and J. Shoham. 1999. Targeting the vulnerable: A review of the necessity and feasibility of targeting vulnerable households. *Disasters* 23 (4): 359–72.

Jezequel, J.-H., and A. Yzebe. 2005. Niger: The sacrificial victims of development. *Messages MSF* 137: 3–4.

Kapp, C. 2005. As Niger's emergency eases, another crisis looms. *The Lancet* 366: 1065–66.

Kokere, S. 2006. *Anthropometric and Nutritional Survey, Tahoua and Illela Districts, Tahoua Region. Summary Report. 2nd–12th December 2005.* Concern.

May, J. F., S. Harouna, and J.-P. Guengant. 2004. *Nourrir, Eduquer et Soigner Tous les Nigeriens, La Démographie en Perspective.* Département du Developpement Humain, Région Afrique, Banque Mondiale.

Miles, M. B., and A. M. Huberman. 1994. *Qualitative Data Analysis.* Thousand Oaks, CA: Sage.

Mousseau, F., and A. Mittal. 2006. *Sahel: A Prisoner of Starvation? A Case Study of the 2005 Food Crisis in Niger.* Oakland, CA: The Oakland Institute.

Rubin, V. 2006. The humanitarian-development debate and chronic vulnerability: Lessons from Niger. *Humanitarian Exchange* 33: 22–24.

SPHERE. 2004. *The Sphere Project: Humanitarian Charter and Minimum Standards in Disaster Response (Revised Edition).* Oxford: Oxfam GB.

Tsai, T.C. 2010. Food crisis no longer taboo in Niger. *The Lancet.* 375 (April 3): 1151–1152.

UNDP. 2006. *UN Human Development Report.* United Nations.

Valid International. 2006. *Community-Based Therapeutic Care (CTC): A Field Manual.* Oxford: Valid International.

World Bank. 2002. *Niger Poverty Reduction Strategy Paper and Joint Staff Assessment.* World Bank 23483-NIR.

World Health Organization. 2001. *The Optimal Duration of Exclusive Breastfeeding: Report of an Expert Consultation.* WHO WHO/FCH/CAH/01.24.

Young, H., and S. Jaspars. 1995a. Malnutrition and poverty in the early stages of famine: North Darfur, 1988–90. *Disasters* 19: 198–215.

———. 1995b. Nutritional assessments, food security and famine. *Disasters* 19: 26–36.

———. 2006. *Meaning and Measurement of Acute Malnutrition, A Primer for Decision-Makers. Network Paper No 56, Humanitarian Practice Network.* London: Overseas Development Institute.

Nutritional Factors in Growth and Disease

• 8 •
Growth, Morbidity, and Mortality in Antiquity
A Case Study from Imperial Rome

T. Prowse, S. Saunders, C. Fitzgerald, L. Bondioli,
and R. Macchiarelli

Introduction

Studies of childhood growth and development in past populations have explained observed patterns of growth faltering, morbidity, and mortality as the result of complex interactions between nutrition and infection (e.g., Hummert and Van Gerven 1983; Goodman et al. 1984; Owsley and Jantz 1985; Mensforth 1985; Lovejoy, Russell, and Harrison 1990; Lewis 2002). It is also recognized that some variables influencing growth and development may not be visible in the skeletal record, such as chronic diarrhea and acute respiratory infections (King and Ulizjaszek 1999). Childhood morbidity and mortality has traditionally been used as an indicator of the overall adaptation of a population to its environment; however, there has been a growing trend in biological anthropology to investigate the lives of children as the primary focus (e.g., Harlow and Laurence 2002; Hewlett and Lamb 2005; Panter-Brick 1998; Perry 2005; Rawson 2003. For a review of the literature see Lewis 2007). While there have been efforts in biological anthropology to compare growth patterns of some prehistoric and historic populations and studies of patterns of infant and young child feeding (IYCF), there is a need to integrate a variety of approaches and types of data in childhood research. Thus, it is important to use as many (and varied) lines of evidence as possible when attempting to evaluate the relative significance of factors affecting childhood growth, nutrition, and health in the past.

This chapter presents a skeletal sample from the Imperial Roman necropolis of Isola Sacra, Italy (ca. first to third centuries CE) as a case study, drawing together multiple lines of evidence from archaeological, historical, skeletal, and dental data to develop a comprehensive picture of the factors affecting childhood development during Roman times. Literary sources from the Roman period provide contextual information on infant- and child-rearing practices from the time of birth, including recommendations on how infants should be breastfed and the diet of children after the weaning process is complete. Microscopic evidence from teeth give us a precise indication of the occurrence and timing of "insults" that are used in tandem with long bone measurements to explore skeletal indications of growth faltering, possibly related to undernutrition and/or disease. Finally, isotopic data provide information on the pattern and timing of weaning and the subsequent diets of young children.

Our observations offer physical evidence of the behavior and attitudes of Romans towards infants and children. It is estimated that nearly 30 percent of infants died before reaching their first birthday (Parkin 1992; Rawson 2003), and Roman period beliefs about infant and child rearing may have compromised their ability to stay alive. While we recognize that a sample of juvenile skeletons is a biased mortality sample (Saunders and Hoppa 2003; Wood et al. 1992), a multifaceted skeletal analysis offers a feasible picture of the effects of biocultural factors on juvenile health and survival in the Imperial Roman period. The integrated analysis of indicators of stress, infant and childhood diet, and long bone development can help to address issues of hidden "frailty" among subgroups in a population (Wood et al. 1992; Wright and Yoder 2003).

The Isola Sacra Skeletal Sample

The necropolis of Isola Sacra is located approximately 23 km southwest of Rome, extending 1.5 km between the port cities of Ostia and Portus Romae (Figure 8.1). The necropolis was used by the inhabitants of Portus Romae between the first and third centuries CE. During the fourth to sixth centuries CE, activity at Portus Romae declined and the necropolis was eventually covered by sand (Baldassarre 1978). The large monumental tombs were first excavated in the 1920s and 1940s by Calza (1940). During the 1970s and 1980s, work at the site focused on the restoration of the monumental tombs, excavation of the areas between the tombs, and recovery of the human skeletal remains reinterred by Calza at the end of his excavations (Angelucci et al. 1990; Baldassarre 1984; Baldassarre et al. 1985; Baldassarre 1990). In addition to large communal tombs, a wide variety of other burial structures were used in the necropolis, including *cappuccina* burials, libation burials, and simple pit burials (see Angelucci et al. 1990). Many of the infants and children were interred in amphorae burials, consisting of the broken pieces of large storage vessels (amphorae). Inscriptions and sculptural reliefs found on some of the monumental tombs at Isola Sacra suggest that many of the people from Portus Romae were traders, merchants, ship owners, and workmen (Toynbee 1941). Inscriptional evidence from the Roman

Growth, Morbidity, and Mortality in Imperial Rome • 175

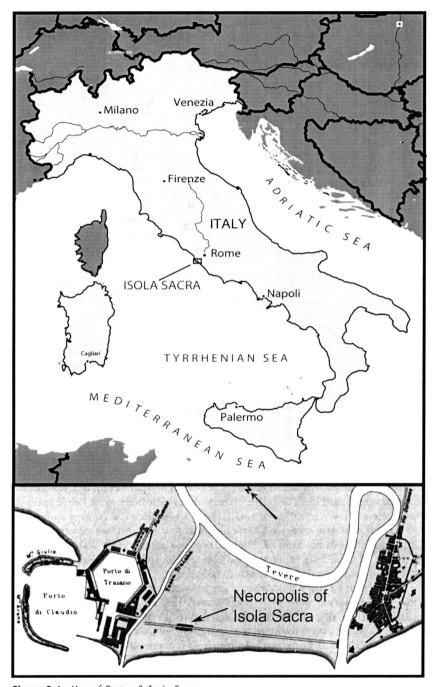

Figure 8.1. Map of Portus & Isola Sacra.

period does not refer to a local aristocracy at Portus Romae, so these people likely represent the middle class of Roman society (Garnsey 1999b).

The total number of individuals in the Isola Sacra sample is estimated to be approximately two thousand (Sperduti 1995). Many of these are commingled remains from Calza's (1940) early excavations, but around eight hundred skeletons have been individually catalogued and analyzed. One of the unique aspects of this skeletal collection is the presence of a large number of infant, child, and adolescent skeletons (n=334). Infants (i.e., those less than one year of age) represent nearly 20 percent of the subadult sample. For the macroscopic analyses, a number of methods were used to determine age-at-death: development and eruption of the deciduous and permanent dentition, development of the temporal and occipital bones, development and fusion of the epiphyses, and maximum long bone diaphyseal length (Sperduti 1995), but dental development was taken as the primary age indicator as long as developing teeth were present for assessment (Saunders 2008).

Infant and Childhood Feeding Practices during the Imperial Roman Period

Historical evidence for attitudes towards breastfeeding and weaning in the Roman period comes primarily from the medical treatises of Soranus and Galen (both from the second century CE) (Galen 1951; Soranus of Ephesus 1956), as well as the fourth-century physician Oribasius (Oribasius 1933). There is also indirect evidence from wet-nursing contracts from Roman Egypt (first to fourth centuries CE) that provide information on the recommended duration of breastfeeding by wet nurses and the appropriate weanling diet (Lefkowitz and Fant 1982). It is important, however, to bear in mind that these Roman period documents were written for a literate, elite segment of Roman society who could afford the choice of hiring a wet nurse if the mother was unable, or unwilling, to breastfeed. Further, these documents were not intended to describe *typical* breastfeeding practices, but rather to prescribe the *best* standard of practice according to medical knowledge at the time, so they should be viewed as the broad context within which the skeletal and dental evidence can be investigated (Prowse et al. 2008). In some instances, recommendations by these authors were clearly detrimental to the survival of infants, such as the advice by Soranus and Oribasius that colostrum should not be fed to newborns because it was thick and difficult to digest (Harlow and Laurence 2002; Lascaratos and Poulakou-Rebelakou 2003). If this advice was followed, it could have deprived newborns of maternal antibodies essential in providing passive immunity to combat early childhood diseases. Extensive swaddling of the infant was also recommended by Soranus from birth up to forty to sixty days in order to ensure that the infant developed into a "well-formed" human being (Harlow and Laurence 2002; Soranus of Ephesus 1956).

Garnsey (1991) suggests that there are two particularly vulnerable periods during the lives of Roman infants. The first occurred around three months of age when infants would have first been exposed to foods other than breast milk, providing an opportunity for exposure to new pathogens. The second occurred when the supply of breast milk was gradually removed and nutritionally inadequate complementary foods were supplied, usually around nine months of age. The pattern and timing of breastfeeding and weaning clearly have an impact on the health and survivorship of infants and young children. Studies of modern human populations have shown that protein is particularly important during the weaning process when the protein-rich breast milk is being removed from the infant diet, because infants and children need more protein by body weight than adults for growth and development (Walker 1990).

It is generally accepted that the ancient medical writers advised a transitional feeding schedule beginning around six months of age and weaning between two to three years of age (Fildes 1986; Lascaratos and Poulakou-Rebelakou 2003; Rawson 2003). These schedules were not precise, but rather were related to stages of development, for example, the period when an infant's first deciduous teeth started to erupt (Garnsey 1991). Wet-nursing contracts from Roman Egypt stipulated that breastfeeding should continue for six months, followed by cow's milk up to eighteen months of age (Lefkowitz and Fant 1982). According to these sources, the first weaning foods consisted of cereals and/or bread softened with milk, sweet wine, or honey wine (Garnsey 1999a).

There is very little evidence from historical sources about the diets of children after weaning was completed. There are sporadic references in ancient texts that identify eggs, porridge, and the shoots of figs as foods recommended for young children (Garnsey 1999a). Galen recommended that the diet of adolescents reflect the activities they would undertake later in life, so an athlete would have a diet rich in pork, beans, and leavened bread (Fidanza 1979). Greco-Roman medical writers described appropriate health regimes for young girls, primarily intended to restrict sexual development until an "appropriate" age for marriage; it was suggested that young females should limit the amount of food consumed, restrict their intake of meat and wine, and engage in work and exercise (Garnsey 1999a). Garnsey (1999a) suggests that inadequate nutrition and disease were prevalent among children under the age of five in the Greco-Roman world.

Stress during Infancy

Wilson Bands

The prevalence of enamel defects, which are caused by disruptions to enamel formation during tooth growth, have long been used in anthropology as proxy measures of the health status and relative adaptive success of past populations (see Hillson 2008, for a review). These visible imperfections in tooth structure occur when the normal process of enamel formation is interrupted or perturbed by some form of external stress. These dental defects represent episodic morbidity events that were survived, because the individual had to live long enough after the stress event(s) in order for the defect(s) to appear in the tooth. The total number of specific conditions that have been associated with defects in clinical studies are too lengthy to list in full here, but they include: intrauterine undernutrition (Grahnén et al. 1972; Norén, Magnusson, and Grahnén 1978), low birth weight (Norén 1983; Norén et al. 1984; Seow 1992), sudden infant death syndrome (Teivens et al. 1996), celiac disease (Horvath and Mehta 2000; Mariani et al. 1994), exposure to polychlorinated biphenyls (PCB) (Jan et al. 2007), cystic fibrosis (Azevedo, Feijo, and Bezerra 2006) and others, as well as many common childhood febrile diseases (Sarnat and Schour 1941; Goodman and Rose, 1990). The weaning process has also been implicated in defect formation in humans and in other mammals (e.g., Franz-Odendaal, Lee-Thorp, and Chinsamy 2003; Matee et al. 1994), and this is important because a number of anthropological studies have tried to relate spikes in population-wide enamel defect prevalence with the initial introduction of complementary foods in the early stages of transitional feeding (e.g., Bermúdez de Castro and Perez 1995; Ensor and Irish 1995; Goodman et al. 1984; Hutchinson and Larsen 1988; Hutchinson and Larsen 1990; Moggi-Cecchi, Pacciani, and Pinto-Cisternas 1994; Simpson 1999; Taji et al. 2000). Most studies have focused on the macroscopic examination of the external manifestation of enamel defects, commonly called linear enamel hypoplasias (LEH). However, in the last decade it has been demonstrated that: 1) the traditional approaches to assigning chronology to the formation of LEH are flawed and inaccurate; 2) traditional macroscopic observation methodologies are imprecise; and 3) depending on the tooth type, some 15 percent to 50 percent of the period of enamel growth is hidden within the cusp and is therefore unable to record stress on the tooth surface in the form of LEH (e.g., Dean and Reid 2001; FitzGerald and Rose 2008; FitzGerald and Saunders 2005; Hillson 1996; Hillson, Antoine, and Dean 1999; Hillson 2005; Hillson and Bond 1997; King, Hillson, and Humphrey 2002; King, Humphrey, and Hillson 2005; Reid, Hillson, and Dean 2000; Reid and Dean 2000). For these reasons we prefer to analyze the *internal* manifestations of the enamel disruptions that appear on the surface (as LEH), variously known as "accentuated striae of Retzius," "pathological striae" or "Wilson bands," the latter being the term that we favor.

Wilson bands are prominent dark striae that are visible for most of their length in the enamel mantle (see FitzGerald and Saunders 2005, Figure 8.1, for a micrograph image of a Wilson band). Wilson bands cross-cut enamel prisms, crystallite rods, many millions of which constitute the structure of enamel. Individual prisms are the product of enamel-producing cells called ameloblasts, which secrete a liquid enamel matrix from their distal ends that immediately begins to mineralize. Secretion starts at what will become the enamel dentine junction (or EDJ) and finishes when the ameloblast reaches the future occlusal surface, where it eventually dies. This means that, unlike bone, enamel cannot remodel and leaves a permanent record of development that is unchanged throughout the existence of the tooth. Wilson bands form when secretion from the whole sheet of functioning ameloblasts is slowed or halted by the effect of the stressor. When secretion resumes, a "kink" or bend is left in each prism forming at the time of the disruption and the cumulative effect in longitudinal section is a band- or striae-like appearance running through the enamel mantle. A Wilson band therefore marks a point of contemporaneity and it represents the external profile of the crown (in three dimensions) at the time of stress.

Utilizing Wilson bands instead of LEH to analyze stress events overcomes the problems mentioned earlier. In particular, it is possible to assign a very accurate chronology to Wilson band formation. The process of ameloblast secretion is under the control of one of the body's central physiological regulators, the circadian or diurnal rhythm. Both the speed of secretion and chemical makeup of the prism varies in a regular periodicity over the course of a day and this is seen under a microscope in polarized light as alternating dark and light bands all along the prism. Since cross striations (which consist of one dark and one light band together) mark a daily passage of time, counting all of them on one prism will therefore yield the number of days taken to form that prism. In addition, there are other microstructures that form with regular periodicity (such as the so-called brown stria of Retzius). However, not only is it possible to determine elapsed time between dental development events, but there is a particular microstructure in enamel that uniquely allows for the assignment of actual chronological ages to these events (as opposed to "biological ages" that some other age-estimation techniques can yield).[1] This is the structure called the *neonatal line,* which forms during the birthing process. It is, in effect, a particularly marked Wilson band that is present in any tooth that is forming perinatally, which in the human dentition are first permanent molars and all of the deciduous teeth. The neonatal line allows timing to be "zeroed" to birth and even if one is not present in the tooth being analyzed, it is possible to cross-reference common structures between teeth and to move from one tooth of known age with a neonatal line, to other teeth in the same dentition without neonatal lines. Using microstructural growth markers of enamel in this way is called *odontochronology,* and very accurate estimates of development events, like morbidity episodes marked by Wilson bands, can be established using this method (see FitzGerald and Saunders 2005 for an example of this method).

The analysis of enamel defects using Wilson bands was undertaken using deciduous teeth in the Isola Sacra sample. Deciduous teeth begin to develop early in fetal life, approximately thirteen to sixteen weeks after fertilization, and the last crowns to develop are the maxillary second molars, which are complete around eleven months after birth (Lunt and Law 1974). Since Wilson bands can only be seen in teeth that are still developing, this means that the window available for the analysis of enamel defects in this sample is restricted to within the first year of life.

Two hundred and seventy-four teeth from 127 Isola Sacra subadults were examined, with ages ranging from birth to approximately thirteen years of age, with a mean age-at-death of 4.2 years (FitzGerald et al. 2006). The teeth were thin-sectioned to produce one or more longitudinal bucco-lingual sections, approximately 70–150 μm thick, taken from the midsection of each tooth. The prepared slides were observed under polarized light, and images were captured with a Polaroid DMC digital video camera attached to the microscope and exported into Adobe Photoshop, which was used to create montages from separate images of each tooth. Data were collected from these images using SigmaScan Pro software from SPSS, Inc. A detailed description of the method for assigning the age of formation to each Wilson band is outlined in FitzGerald and Saunders (2005). Briefly, the neonatal line was identified in each section and a photomicrograph that included both the Wilson band and the neonatal line was taken. The path of one prism running between these two microstructures was traced on the image, and cross striations were counted and/or calculated along the prism, to produce a chronology of prism formation between birth and Wilson band formation.

Long Bone Growth

Skeletal growth in archaeological samples has been studied as an indicator of the overall well-being of past populations and, in particular, childhood living conditions (Johnston and Zimmerman 1989; Saunders and Hoppa 2003). The pattern of growth from infancy through adulthood is normally continuous, although there are two periods of accelerated growth: one during the first year of life and a second during adolescence (the adolescent growth spurt) (Bogin 1999; Larsen 1997). Growth can be affected by a wide variety of factors, but it is generally acknowledged that the interaction of undernutrition and infectious disease has the greatest impact on growth (Humphrey 2000; Larsen 1997; Saunders 2008). It is also well established that most deficiencies in stature during growth are due to retardation of leg growth (diaphyseal growth) during infancy and childhood when undernutrition and disease have significant influences on those living in poverty or poor circumstances (Bogin 2001).

Numerous studies have examined skeletal growth in archaeological samples and have found that the pattern of growth in past groups is generally similar to that of modern populations (see Humphrey 2000; Larsen 1997; Saunders 2008 for reviews). It is inferred that any deviation from this pattern of growth can be attributed to some kind of "chronic stressor" (e.g., malnutrition, undernutrition, and/or disease) acting on the individual (Johnston and Zimmerman 1989; Larsen 1997; Saunders 2008), much like those known to cause enamel defects. Comparative analysis of growth in skeletal samples involves the measurement of diaphyseal lengths of long bones in relation to estimated age-at-death, which is usually based on dental development. This type of comparison, however, assumes that the patterns of dental development are similar in the skeletal samples under study. Some have challenged the oft-cited assumption, however, that dental development is less susceptible to environmental, nutritional, and genetic factors (see Saunders 2008).

Femoral diaphyseal length data from children ranging in age from birth to thirteen years in the Isola Sacra sample (n=248) are compared to previously published data from a nineteenth-century skeletal sample from Belleville, Ontario (n=142)

(Saunders, Hoppa, and Southern 1993; Sperduti et al. 1997). The St. Thomas' sample was previously recognized as being close to modern American standards of growth of living children (Saunders, Hoppa, and Southern 1993). Diaphyseal lengths were measured to the nearest whole millimeter. A comparative skeletal growth profile was created by plotting femoral diaphyseal lengths against estimates of chronological age based on tooth formation. When available, the left side was used, so that each individual was represented only once in the skeletal growth profile (after Saunders, Hoppa, and Southern 1993).

Stable Isotope Analysis

Breastfeeding and Weaning

The stable isotopes of nitrogen ($\delta^{15}N$) and carbon ($\delta^{13}C$) in bone collagen are routinely used to study IYCF patterns in skeletal samples (e.g., Clayton, Sealy, and Pfeiffer 2006; Dupras, Schwarcz, and Fairgrieve 2001; Herring, Saunders, and Katzenberg 1998; Katzenberg, Herring, and Saunders 1996; Mays, Richards, and Fuller 2002; Ogrinc and Budja 2005; Richards, Mays, and Fuller 2002; Schurr 1997; Schurr 1998; Schurr and Powell 2005; Turner et al. 2007; White and Schwarcz 1994). These isotopic studies help reveal the relationships among maternal behavior, infant and young child feeding practices, morbidity, and mortality in past populations (Katzenberg, Herring, and Saunders 1996; Dupras, this volume). Sellen (this volume) defines *weaning* as the termination of breastfeeding and *transitional feeding* as the phase during which both breast milk and complementary foods are provided. *Complementary foods* are those obtained and processed for infants over the age of six months. These definitions are different than those commonly used in isotopic studies of weaning, but we use them here to relate this study to the broader biocultural research on breastfeeding and weaning in past and present populations.

Research on marine and terrestrial organisms indicate a trophic level shift for both $\delta^{15}N$ and $\delta^{13}C$ values, such that the tissues of consumers are enriched by approximately 3‰ (per mil) and 1‰, respectively, relative to the diet consumed (Deniro and Epstein 1978; Fuller et al. 2006; Minagawa and Wada 1984; Schoeninger 1985; Schoeninger and Deniro 1984). Breastfeeding infants have similarly enriched $\delta^{15}N$ and $\delta^{13}C$ levels, because the infants are consuming the mothers' tissues (Fogel et al. 1997; Fuller et al. 2006). There is a general increase in isotopic values with increasing age-at-death in infants and children while they are breastfeeding, reaching a peak value at roughly one trophic level above adult values, followed by a progressive decline in isotopic values over time as breast milk is gradually removed from the diet. This pattern is attributed to the process of transitional feeding and weaning, with $\delta^{15}N$ and $\delta^{13}C$ values ultimately reaching levels representative of a postweaning childhood diet. One of the uncertainties in IYCF studies is the rate of bone deposition and turnover in growing infants; thus it is not possible to determine the lag time between complete cessation of breastfeeding and the associated shift in isotopic levels (Herring, Saunders, and Katzenberg 1998). Beardsworth, Eyre, and Dickson (1990) found that biochemical indicators of collagen resorption were 12.4–14.4 times higher in children (aged two to fifteen years) than adults, reflecting a much greater rate of bone formation and turnover. Since the samples come from infants who did not survive, it is not clear what factors attributed to cause of death. Infant mortality may be related to the exposure of infants to new pathogens with the introduction of complementary foods and other sources of infection.

Thirty-seven rib samples were analyzed, based on the availability of sufficient bone for stable isotope analysis (approximately 3–5 grams) (Prowse et al. 2008). The preparation of bone collagen for isotopic analysis followed the methods of Longin (1971) and Chisholm, Nelson, and Schwarcz (1982). Approximately 2–3 mg (carbon) and 9–13 mg (nitrogen) of the dried collagen was reacted in separate tubes with CuO (at 550 °C for 2.5 hours, oxidizing the collagen to CO_2, N_2, and H_2O). The prepared collagen samples were analyzed using a VG SIRA 10 Series II mass spectrometer at McMaster University.

The data are presented in the d-notation:

$$\delta^{15}N = \{(R_x/R_s) - 1\} \times 1000‰ \text{ (per mil)}$$

where R = $^{15}N/^{14}N$; x = sample; s = standard (and similarly for $\delta^{13}C$). The standards are PDB (Peedee Belemnite) for $\delta^{13}C$ and atmospheric N_2 (AIR) for $\delta^{15}N$. The precision of analysis is ~ 0.1‰ for $\delta^{13}C$ and ~ 0.2‰ for $\delta^{15}N$.

Results

Wilson Bands

Fifty individuals (50/127 = 39.4 percent) have one or more Wilson bands in at least one tooth (FitzGerald et al. 2006). If the data are analyzed by tooth, 64 of the 274 teeth (23.4 percent) have at least one Wilson band (FitzGerald et al. 2006). A total of 447 Wilson bands were observed on the 64 affected teeth, averaging 7.0 Wilson bands per affected tooth. A series of adjustments were made to the raw data to correct for factors that will invalidate the calculation of prevalence (the number of cases of a condition ÷ total population at risk). These included adjustments to account for tooth sampling variation and infant and child death, and an adjustment to record only the highest level of stress registering in any tooth type at any moment of development, resulting in a distribution curve we call the maximum prevalence (MAP) (for a more detailed discussion of the adjustments identified below see FitzGerald et al. 2006). A final adjustment was made to account for possible sampling bias in our cemetery population, as well as other factors including artifacts of analysis (such as forcing continuous events into monthly classes), making our MAP distribution likely unrealistically "spiky." Therefore, we calculated smoothed maximum prevalence (SMAP), the maximum prevalence distribution drawn as a smoothed curve based on a trend line produced using higher-order (sixth) polynomial regression.

Figure 8.2 shows the maximum prevalence (MAP) and smoothed maximum prevalence (SMAP) distribution of Wilson bands in each month, reflecting the changes discussed above. As can be seen in this figure, the prevalence of stress events increases dramatically in the second month and reaches a maximum of 80 percent in the fifth month. This high prevalence continues until the beginning of month nine, after which there is a decline towards the end of the first year. Months eleven and twelve are not included in the diagram, since the number of teeth and tooth types remaining in the sample by this time are too few to accurately represent the true prevalence of Wilson bands in this sample. Further, the SMAP curve for months ten and eleven also underestimates the true prevalence of Wilson bands, because the bands that occur in these months are located toward the cervix of the only two tooth types (canines and second molars) left in the sample; that is, the other deciduous teeth (incisors and first molars) have finished forming, so no more stress events can be recorded in those teeth.

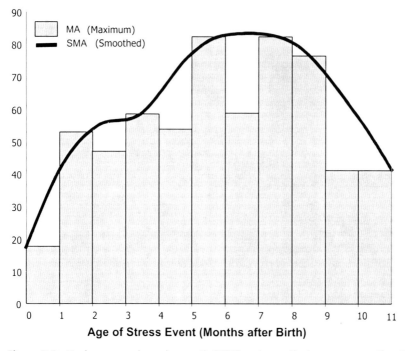

Figure 8.2. Maximum prevalence by month (MAP) and smoothed curve, suggesting the shape of the 'true' prevalence distribution (SMAP).

Long Bone Length

We plotted femoral diaphyseal length against dental age (using the formation standards provided by Moorrees et al. 1963a, 1963b) for both the Isola Sacra sample and the nineteenth-century skeletal sample from Belleville, Ontario, the St. Thomas sample (Figure 8.3). This figure shows the two samples overlapping up to about eight years of age. After eight years there is a divergence with the Isola Sacra sample often having smaller-for-age individuals.

One thing that must be remembered when investigating growth in archaeological samples is that the data are derived from nonsurvivors; therefore the results do not represent truly cross-sectional or longitudinal growth studies based on living populations (Saunders 2008; Saunders and Hoppa 1993). Consequently, these data suggest the following interpretation of the results. Since almost all of the children in the St. Thomas' mortality sample (except for some outliers) reached statures comparable to healthy, living children, morbidity and mortality must have been mainly due to acute causes such as rapid infections (which is supported by the documentary literature). Similarly, the Isola Sacra infants and young children appear to have died of mainly acute causes (i.e., there is not much evidence of chronic growth stress at least from

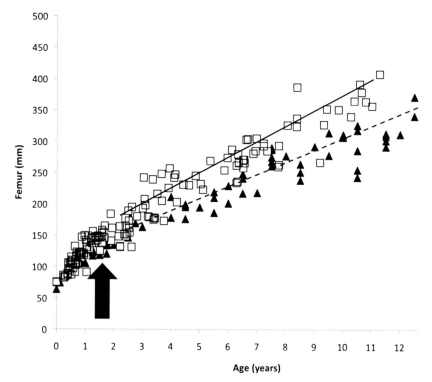

Figure 8.3. Isola Sacra vs. Belleville Diaphyseal Lengths.

growth of the long bones), but by the time mid- to late childhood was reached, some of the children were showing the effects of chronic stress affecting growth, and this became manifest in their statures as well as the statures of some of the adults.

In order to control for the possible confounding factor that there may be population differences in the rates of dental development of the two samples (which was used as the proxy estimation of chronological age-at-death), Saunders (2008) compared the frequency distribution of formation stages of three pairs of teeth from the Isola Sacra sample and St. Thomas samples. Results demonstrated that there are no systematic differences between the two samples in terms of the rate and pattern of tooth formation. The absence of any variation in patterns of dental development between the two samples negates any expectations of microlevel genetic differences in dental development between the two samples or differences due to environmental factors that might affect dental formation. This suggests that comparisons of skeletal growth between the two samples, using the same standards for dental formation (i.e., Moorrees et al. 1963a; 1963b), are appropriate (Saunders 2008).

The absence of substantive differences in diaphyseal size between the Isola Sacra and the St. Thomas samples is in strong contrast to other growth studies of archaeological samples (see for example, Hoppa 1992; Humphrey and Stringer 2002; Saunders, Hoppa, and Southern 1993). In addition, a more recent study by Wood (2004) has identified a high prevalence of childhood rickets in the Isola Sacra sample, which is known to cause chronic growth stunting. However, children experiencing fluctuating levels of Vitamin D exposure (90 percent of which usually comes from exposure to sunlight) may undergo catch-up growth, and the observation of bowing of the long bones as a diagnostic feature of the disease actually indicates that the child was healthy enough to attempt walking (Ortner and Mays 1998). It is perhaps not surprising then that the compromising of diaphyseal growth is not clearly manifest in the mortality sample until later childhood.

Stable Isotope Analysis

The $\delta^{15}N$ data are plotted in Figure 8.4, with the average adult female mean indicated by the solid line (10.6 ± 1.1‰) (Prowse et al. 2004). The $\delta^{15}N$ results range from 8.7 to 16.1‰ (mean = 12.4 ± 1.9‰). Individuals younger than two years of age-at-death tend to have higher $\delta^{15}N$ values (mean = 13.5 ± 1.5‰) than those older than the age of two (mean = 11.2 ± 1.4‰). A *t*-test confirms that $\delta^{15}N$ values are significantly higher among individuals younger than two years of age (P < 0.01). The highest $\delta^{15}N$ values are among individuals under the age of one year and there is relatively little variation in isotopic values among individuals in this age range (approximately 1‰). There is, however, considerably more variability in the $\delta^{15}N$ data among individuals between one and two years of age-at-death. The maximum offset in $\delta^{15}N$ values with respect to adult values is approximately 4‰, which is slightly larger than the expected trophic level shift of 3‰ associated with the removal of breast milk from the infant diet. This is not surprising, since the $\delta^{15}N$ values of breastfeeding

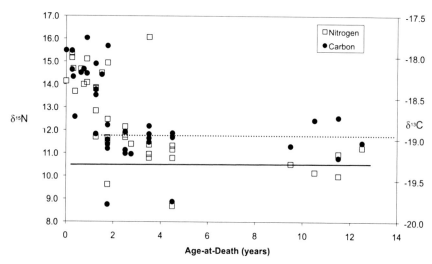

Figure 8.4. Scatter plot of the Isola Sacra rib data (n=37), showing $\delta^{15}N$ and $\delta^{13}C$ versus estimated age-at-death. The solid horizontal line represents the adult female $\delta^{15}N$ mean (10.6 ‰); the dotted horizontal line represents the adult female $\delta^{13}C$ mean (−18.9‰).

infants is affected, in part, by the maternal diet. If the mother had been consuming foods at higher trophic levels (e.g., fish or seafood) so that her $\delta^{15}N$ values were higher than average, then her infant's $\delta^{15}N$ values would also reflect this, in addition to the trophic level enrichment due to breast milk.

The $\delta^{13}C$ results are also shown in Figure 8.4, and range from −17.8 to −19.8‰ (mean = −18.7 ± 0.5‰). There is a cluster of more positive $\delta^{13}C$ values under the age of two, but the data are not as divergent as seen in the nitrogen data, with a mean value of −18.5 ± 0.5‰ for individuals younger than two years and a mean value of −19.1 ± 0.2‰ for those older than two years at the time of death. A *t*-test indicates that $\delta^{13}C$ values are significantly more negative among children older than two years of age ($P < 0.01$). The $\delta^{13}C$ data show a smaller, but definite decrease in the isotopic levels, approximately 1‰ (from −18 to −19‰), which corresponds to an expected trophic level effect for carbon. As with the nitrogen data, there is no clear pattern in $\delta^{13}C$ data among the youngest infants (i.e., < one year). Between birth and twelve to eighteen months, the isotopic values are scattered, although the data do appear to decrease between one and two-and-a-half years of age, after which the values approximate (but are generally below) the adult mean (indicated by the dotted horizontal line).

Discussion and Conclusions

Garnsey (1999a) hypothesizes that there was a high incidence of morbidity and mortality in children under the age of five in Roman society, which he attributes in part to infant feeding practices. It is clear that Romans were familiar with, if not accustomed to, high infant mortality in the first few weeks of life. The actual naming of the infant did not occur until eight (females) or nine (males) days after birth, since it was considered pointless to name a child until it had survived the first week of life (Harlow and Laurence 2002; Wiedemann 1989). Other recommended childcare practices may have also contributed to morbidity and mortality in Roman infants. Soranus recommended that newborns should not be fed until two days after birth and that colostrum was unhealthy for the infant (Soranus of Ephesus 1956). Tight swaddling of the infant was also recommended for the first few months of life in order to develop a well-formed body (Harlow and Laurence 2002; Rawson 2003). Modern studies have demonstrated the calming effect of intermittent swaddling on infants (e.g., Gerard, Harris, and Thach 2002); however, prolonged periods of swaddling may limit infants' exposure to sunlight (promoting Vitamin D deficiency) and could also lead to respiratory and biomechanical problems (Yurdakok, Yavus, and Taylor 1990; Kutlu et al. 1992).

If the documented prescriptions for infant feeding were actually incorporated into cultural norms found among the populace of Imperial Rome, then the agreement between the Wilson band histological data in this sample and reconstructed behavioral descriptions are likely not chance findings. The observed prevalence of Wilson bands seen in Figure 8.2 can be divided into several episodes. After an initial dramatic rise, the curve reaches a maximum at around two months that is maintained through month five, and then a second, higher level is reached around month six, which continues through month nine. This pattern appears broadly consonant with both general expectations for risk periods during the first year and for descriptions given for the classical Roman period. However, it must also be borne in mind that morbidity events other than those associated with the weaning process are also involved. What the SMAP curve reflects is the general risk to health experienced during infancy, some contribution to which undoubtedly arose from the process of introducing complementary foods and withdrawing breast milk, but some portion of which may be also attributable to other morbidity events in the lives of these infants.

The isotopic analysis of the rib samples indicates a clear pattern of higher $\delta^{13}C$ and $\delta^{15}N$ values in the youngest members of the skeletal sample, with individuals less than two years of age possessing higher $\delta^{13}C$ and $\delta^{15}N$ values than those above two years of age. There is slight variability in the $\delta^{15}N$ values prior to one year of age; however, the greatest range of variation occurs among infants aged one to two years. However, we do not yet fully understand the rate of bone formation and turnover among infants, so it is not clear how much time is required for the bones of an infant to register the removal of breast milk from the diet. If we detect a trend of decreasing isotopic values starting around twelve months, this suggests that the process began somewhat earlier than this, which is consistent with the histological evidence for an extended period of "stress" between six to nine months of age. One thing that cannot be determined from the isotopic data is whether the complementary foods were nutritionally adequate. What can be said is that the pattern of breastfeeding and weaning is generally consistent with descriptions in the ancient sources, and this may have had serious implications for infant and childhood health, as passive immunity is lost with the removal of breast milk and risk of infectious disease increases with the introduction of complementary foods (Katzenberg 2008).

One possible way to study the rate of bone formation is to compare different age estimators for the infant (i.e., dental development versus long bone length). If the age estimates differ significantly, with long bone lengths indicating a younger age, then it is possible that the infants were small for their age; a fact which could be related to rate of modeling, or to the deviation of some individuals from the main growth trend. However, the comparison of femur lengths among the Isola Sacra children to those from an early nineteenth-century Canadian sample showed that growth and development were similar in both samples until approximately eight years of age (Saunders et al. 2000; Saunders 2008). This implies that the Isola Sacra infants and young children were not small for their age.

This study demonstrates that multiple lines of evidence can be integrated to investigate growth, morbidity, and mortality in skeletal samples. The written evidence provides useful insight into the possible quality of the diet, and supplies invaluable historical context for comprehending the patterns seen in the skeletal and dental remains. Information concerning cultural attitudes towards breastfeeding and weaning are important to understanding infant and childhood health in antiquity. The histological evidence suggests that infants (under one year of age) were exposed to "stressors" around six to nine months of age. Sellen (this volume) suggests that human infants have not evolved to effectively utilize foods other than breast milk prior to six months of age, and that the human pattern of "early and flexible" weaning is characterized by the introduction of complementary foods after six months when infants are physiologically ready. The histological data support this pattern of early and flexible weaning. The isotopic evidence indicates that the process of transitional feeding was well underway between one and two years of age, and that weaning was completed by two-and-a-half years of age. It is likely that transitional feeding began before the end of the first year of life, but there is a delay in that signal being registered in the infants' bones. Finally, the long bone growth data help put the histological and isotopic data into context, indicating that even though these infants were going through periods of "stress," likely associated with the removal of breast milk and the introduction of complementary foods, their growth was not significantly affected. Since deviations in growth patterns can be attributed to chronic stressors such as malnutrition, undernutrition, and/or disease (see the earlier discussion in this chapter on long bone growth), this suggests that the individual stress events were not chronic. It must be remembered, however, that we are dealing with those infants and children who died, and we may not be able to discern the cause of death. Paleopathological analysis of this skeletal sample may help to shed light on morbidity events in the sample.

Notes

1. *Chronological age* refers to the amount of time one has been alive, whereas *developmental age* refers to the level of skeletal growth and maturity, independent of the time one has been alive.
2. For those interested in more detail about the process of enamel formation and the techniques of utilizing microstructures to establish chronology, see Boyde 1963, 1976, 1989; Dean 1987; Dean et al. 1993; FitzGerald 1998; FitzGerald and Rose 2008; Hillson 1996, 2005.
3. Paired tooth comparisons were (1) mandibular m1 and M1, (2) mandibular m2 and M1, and (3) mandibular M1 and M2.

References

Angelucci, S., I. Baldassarre, I. Bragantini, M. G. Lauro, V. Mannucci, A. Mazzoleni, C. Morselli, and F. Taglietti. 1990. Sepolture e riti nella necropoli dell'Isola Sacra. *Bollettino di Archeologia* 56: 50–113.

Azevedo, T. D., G. C. Feijo, and A. C. Bezerra. 2006. Presence of developmental defects of enamel in cystic fibrosis patients. *Journal of Dentistry for Children* 73: 159–63.

Baldassarre, I. 1978. La Necropoli Dell'Isola Sacra. *Quaderni de 'La Ricerca Scientifica'* 100: 3–20.

———. 1984. Una necropoli imperiale Romana proposta di lettura. *Archeologia e Storia Antica* 6: 141–49.

———. 1990. Nuove ricerche nella necropoli dell'Lsola Sacra. *Archeologia Laziale* 10: 164–72.

Baldassarre, I., I. Bragatanini, A. M. Dolciotti, C. Morselli, F. Taglietti, and M. Taloni. 1985. La necropoli dell'Isola Sacra campagna di scavo 1976–1979. *Quaderni de 'La Ricerca Scientifica'* 112: 261–302.

Beardsworth, L. J., D. R. Eyre, and I. R. Dickson. 1990. Changes with age in the urinary excretion of lysyl- and hydroxylysylpyridinoline, two new markers of bone collagen turnover. *Journal of Bone and Mineral Research* 5: 671–76.

Bermúdez de Castro, J. M. and P. J. Perez. 1995. Enamel hypoplasia in the middle Pleistocene hominids From Atapuerca (Spain). *American Journal of Physical Anthropology* 96: 301–14.

Bogin, B. 1999. *Patterns of Human Growth.* Cambridge: Cambridge University Press.

———. 2001. *The Growth of Humanity.* New York: Wiley-Liss.

Boyde, A. 1963. Estimation of age at death of young human skeletal material from incremental lines in the dental enamel (abstract). Third International Meeting in Forensic Immunology, Medicine, Pathology, and Toxicology (16–24 April 1963). London. Plenary Session 11A.

———. 1976. Amelogenesis and the structure of enamel. In *Scientific Foundations of Dentistry,* ed. B. Cohen and I. R. H. Kramer, 335–52. London: Heinemann.

———. 1989. Enamel. In *Handbook of Microscopic Anatomy,* vol. V/6, *Teeth,* ed. A. Oksche and L. Vollrath, 309–473. Berlin: Springer-Verlag.

Calza, G. 1940. *La Necropoli Del Porto Di Roma Nell'Isola Sacra.* Roma: Instituto di Archeologia e Storia dell'Arte.

Chisholm, B. S., D. E. Nelson, and H. P. Schwarcz. 1982. Stable-carbon isotope ratios as a measure of marine versus terrestrial protein in ancient diets. *Science* 216: 1131–32.

Clayton, F., J. Sealy, and S. Pfeiffer. 2006. Weaning age among foragers at Matjes River Rock Shelter, South Africa, from stable nitrogen and carbon isotope analyses. *American Journal of Physical Anthropology* 129: 311–17.

Dean, M. C. 1987. Growth layers and incremental markings in hard tissues, a review of the literature and some preliminary observations about enamel structure in *Paranthropus boise*. *Journal of Human Evolution* 16: 157–72.

Dean, M. C., A. D. Beynon, D. J. Reid, and D. K. Wittaker. 1993. A longitudinal study of tooth growth in a single individual based on long- and short-period incremental markings in dentine and enamel. *International Journal of Osteoarchaeology* 3: 249–64.

Dean, M. C., and D. J. Reid. 2001. Perikymata spacing and distribution on hominid anterior teeth. *American Journal of Physical Anthropology* 116: 209–15.

Deniro, M. J., and S. Epstein. 1978. Influence of diet on the distribution of carbon isotopes in animals. *Geochimica et Cosmochimica Acta* 42: 495–506.

Dupras, T. L., H. P. Schwarcz, and S. I. Fairgrieve. 2001. Infant feeding and weaning practices in Roman Egypt. *American Journal of Physical Anthropology* 115: 204–12.

Ensor, B. E., and J. D. Irish. 1995. Hypoplastic area method for analyzing dental enamel hypoplasia. *American Journal of Physical Anthropology* 98: 507–17.

Fidanza, F. 1979. Diets and dietary recommendations in ancient Greece and Rome and the school of Salerno. *Progress in Food and Nutrition Science* 3: 79–99.

Fildes, V. 1986. *Breast, Bottles, and Babies: A History of Infant Feeding.* Edinburgh: Edinburgh University Press.

FitzGerald, C. M. 1998. Do enamel microstructures have regular time dependency? Conclusions from the literature and a large-scale study. *Journal of Human Evolution* 35: 371–86.

FitzGerald, C. M., and J. C. Rose. 2008. Reading between the lines: Dental development and subadult age assessment using the microstructural growth markers of teeth. In *The Biological Anthropology of the Human Skeleton,* 2nd ed., ed. M. A. Katzenberg and S. R. Saunders, 237–64. New York: John Wiley & Sons, Inc.

FitzGerald, C. M. and S. R. Saunders. 2005. Test of histological methods of determining chronology of accentuated striae in deciduous teeth. *American Journal of Physical Anthropology* 127: 277–90.

FitzGerald, C. M., S. R. Saunders, R. Macchiarelli, and L. Bondioli. 2006. Health of infants in an imperial Roman skeletal sample: Perspective from dental microstructure. *American Journal of Physical Anthropology* 130: 179–89.

Fogel, M. L., N. Tuross, B. J. Johnson, and G. H. Miller. 1997. Biogeochemical record of ancient humans. *Organic Geochemistry* 27: 275–87.

Franz-Odendaal, T. A., J. A. Lee-Thorp, and A. Chinsamy. 2003. Insights from stable light isotopes on enamel defects and weaning in Pliocene herbivores. *Journal of Biosciences* 28: 765–73.

Fuller, B. T., T. I. Molleson, D. A. Harris, L. T. Gilmour, and R. E. M. Hedges. 2006. Isotopic evidence for breastfeeding and possible adult dietary differences from late/sub-Roman Britain. *American Journal of Physical Anthropology* 129: 45–54.

Galen. 1951. *Hygiene (De Sanitate Tuenda)*, trans. R. M. Green. Springfield, IL: Charles C. Thomas.

Garnsey, P. 1991. Child rearing in ancient Italy. In *The Family in Italy: From Antiquity to the Present*, ed. D. I. Kertzer and R. P. Saller. New Haven, CT: Yale University Press, 48–65.

———. 1999a. *Food and Society in Classical Antiquity*. Cambridge: Cambridge University Press.

———. 1999b. The People of Isola Sacra. In *Digital Archives of Human Paleobiology, Osteodental Biology of the People of Portus Romae (Necropolis of Isola Sacra, 2nd–3rd Cent. AD)*, vol. 1, *Enamel Microstructure and Developmental Defects of the Primary Dentition*, ed. L. Bondioli and R. Macchiarelli. Rome: Section of Anthropology, National Prehistoric Ethnographic L. Pigorini Museum.

Gerard, C. M., K. A. Harris, and B. T. Thach. 2002. Physiologic studies on swaddling: An ancient child care practice, which may promote the supine position for infant sleep. *The Journal of Pediatrics* 141 (3): 398–404.

Goodman, A. H., G. J. Armelagos, and J. C. Rose. 1984. The chronological distribution of enamel hypoplasias from prehistoric Dickson Mounds populations. *American Journal of Physical Anthropology* 65: 259–66.

Goodman, A. H., J. Lallo, G. J. Armelagos, and J. C. Rose. 1984. Health changes at Dickson Mounds, Illinois (A.D. 950–1300) In *Palaeopathology at the Origins of Agriculture*, ed. M. N. Cohen and G. J. Armelagos, 271–306. New York: Academic Press.

Goodman, A. H. and J. C. Rose. 1990. Assessment of systemic physiological perturbations from dental enamel hypoplasias and associated histological structures. *Yearbook of Physical Anthropology* 33: 59–110.

Grahnén, H., A. K. Holm, S. Sjölin, and B. Magnusson. 1972. Mineralisation defects of the primary teeth in intra-uterine undernutrition *Caries Research* 6: 224–28.

Harlow, M., and R. Laurence. 2002. *Growing Up and Growing Old in Ancient Rome: A Lifecourse Approach*. London: Routledge.

Herring, D. A., S. R. Saunders, and M. A. Katzenberg. 1998. Investigating the weaning process in past populations. *American Journal of Physical Anthropology* 105: 425–39.

Hewlett, B. S., and M. E. Lamb. 2005. *Hunter-Gatherer Childhoods: Evolutionary, Developmental, and Cultural Perspectives*. Piscataway, NJ: Aldine Transaction.

Hillson, S. W. 1996. *Dental Anthropology*. Cambridge: Cambridge University Press.

———. 2005. *Teeth, Second Edition*. Cambridge: Cambridge University Press.

———. 2008. Dental Pathology. In *Biological Anthropology of the Human Skeleton*, 2nd ed., ed. M. A. Katzenberg and S. R. Saunders, 301–40. New York: John Wiley & Sons Inc.

Hillson, S. W., D. M. Antoine, and M. C. Dean. 1999. A detailed developmental study of the defects of dental enamel in a group of post-medieval children From London. In *Proceedings of the 11th International Symposium of Dental Morphology, Aug 26–30, 1998, Oulu, Finland*, ed. J. T. Mayhall and H. Heikkinen, 102–11. Oulu, Finland: Oulu University Press.

Hillson, S. W., and S. Bond. 1997. Relationship of enamel hypoplasia to the pattern of tooth crown growth: A discussion. *American Journal of Physical Anthropology* 104: 89–103.

Hoppa, R. D. 1992. Evaluating human skeletal growth: An Anglo-Saxon example. *International Journal of Osteoarchaeology* 2: 275–88.

Horvath, K., and D. I. Mehta. 2000. Celiac disease—A worldwide problem. *Indian Journal of Pediatrics* 67: 757–63.

Hummert, J. R., and D. P. Van Gerven. 1983. Skeletal growth in a medieval population from Sudanese Nubia. *American Journal of Physical Anthropology* 60: 471–78.

Humphrey, L. T. 2000. Growth studies of past populations: An overview and an example. In *Human Osteology in Archaeology and Forensic Science*, ed. M. Cox and S. A. Mays, 28–38. London: Greenwich Medical Media.

Humphrey, L. T., and C. B. Stringer. 2002. The human cranial remains from Gough's Cave (Somerset England). *Bulletin of the Natural History Museum of London* 58: 153–68.

Hutchinson, D. L., and C. S. Larsen. 1988. Determination of stress episode duration from linear enamel hypoplasias: A case study from St. Catherine's Island, Georgia. *Human Biology* 60: 93–110.

———. 1990. Stress and lifeway changes: The evidence from enamel hypoplasias. In *The Archaeology of Mission Santa Catalina De Guale: 2. Biocultural Interpretations of a Population in Transition*, ed. C. S. Larsen, 50–65. New York: American Museum of Natural History.

Jan, J., E. Sovcikova, A. Kočan, L. Wsolova, and T. Trnovec. 2007. Developmental dental defects in children exposed to PCBs in eastern Slovakia. *Chemosphere* 67: S350–S354.

Johnston, F. E., and L. J. Zimmerman. 1989. Assessment of growth and age in the immature skeleton. In *Reconstruction of Life from the Skeleton*, ed. M. Y. Iscan and K. A. R. Kennedy, 11–22. New York: Alan R. Liss.

Katzenberg, M. A. 2008. Stable isotope analysis: A tool for studying past diet, demography, and life history. In *Biological Anthropology of the Human Skeleton*, 2nd ed., ed. M. A. Katzenberg and S. R. Saunders, 413–42. New York: Wiley-Liss.

Katzenberg, M. A., D. A. Herring, and S. R. Saunders. 1996. Weaning and infant mortality: Evaluating the skeletal evidence. *Yearbook of Physical Anthropology* 39: 177–99.

King, S. E., and S. J. Ulijaszek. 1999. Invisible insults during growth and development: Contemporary theories and past populations. In *Human Growth in the Past: Studies from Bones and Teeth*, ed. R. D. Hoppa and C. M. FitzGerald, 161–82. Cambridge: Cambridge University Press.

King, T., S. W. Hillson, and L. T. Humphrey. 2002. A detailed study of enamel hypoplasia in a post-medieval adolescent of known age and sex. *Archives of Oral Biology* 47: 29–39.

King, T., L. T. Humphrey, and S. W. Hillson. 2005. Linear enamel hypoplasias as indicators of systemic physiological Stress: Evidence from two known age-at-death and sex populations from postmedieval London. *American Journal of Physical Anthropology* 128: 547–59.

Kutlu, A., R. Memik, M. Mutlu, R. Kutlu, and A. Arslan. 1992. Congenital dislocation of the hip and its relation to swaddling used in Turkey. *Journal of Pediatric Orthopedics* 12: 598–602.

Larsen, C. S. 1997. *Bioarchaeology.* Cambridge: Cambridge University Press.
Lascaratos, J., and E. Poulakou-Rebelakou. 2003. Oribasius (fourth century) and early Byzantine perinatal nutrition. *Journal of Pediatric Gastroenterology and Nutrition* 36: 186–89.
Lefkowitz, M. R., and M. B. Fant. 1982. *Women's Life in Greece and Rome.* Toronto: Samuel-Stevens.
Lewis, M. E. 2002. *Urbanisation and Child Heath in Medieval and Post-Medieval England.* Oxford: Archaeopress.
———. 2007. *The Bioarchaeology of Children.* Cambridge: Cambridge University Press.
Longin, R. 1971. New method of collagen extraction for radiocarbon dating. *Nature* 230: 241–42.
Lovejoy, C. O., K. F. Russell, and M. L. Harrison. 1990. Long bone growth velocity in the Libben population. *American Journal of Human Biology* 2: 533–41.
Lunt, R. C., and D. B. Law. 1974. A review of the chronology of eruption of deciduous teeth. *Journal of the American Dental Association* 89: 872–79.
Mariani, P., M. C. Mazzilli, G. Margutti, P. Lionetti, P. Triglione, F. Petronzelli, E. Ferrante, and M. Bonamico. 1994. Coeliac disease, enamel defects and HLA typing. *Acta Paediatrica* 83: 1272–75.
Matee, M., M. van't Hof, S. Maselle, F. Mikx, and W. van Palenstein Helderman. 1994. Nursing caries, linear hypoplasia, and nursing and weaning habits in Tanzanian infants. *Community Dentistry and Oral Epidemiology* 22: 289–93.
Mays, S. A., M. P. Richards, and B. T. Fuller. 2002. Bone stable isotope evidence for infant feeding in mediaeval England. *Antiquity* 76: 654–56.
Mensforth, R. P. 1985. Relative tibia long bone growth in the Libben and Bt-5 prehistoric skeletal populations. *American Journal of Physical Anthropology* 68: 247–62.
Minagawa, M., and E. Wada. 1984. Stepwise enrichment of ^{15}N along food chains: Further evidence and the relation between D^{15}N and animal age. *Geochimica et Cosmochimica Acta* 48: 1135–40.
Moggi-Cecchi, J., E. Pacciani, and J. Pinto-Cisternas. 1994. Enamel hypoplasia and age at weaning in 19th-century Florence, Italy. *American Journal of Physical Anthropology* 93: 299–306.
Moorrees, C. F., E. A. Fanning, and E. E. Hunt Jr. 1963a. Age variation of formation stages for ten permanent teeth. *Journal of Dental Research* 42: 1490–1502.
———.1963b. Formation and resorption of three deciduous teeth in children. *American Journal of Physical Anthropology* 21: 205–13.
Norén, J. G. 1983. Enamel structure in deciduous teeth from low-birth-weight infants. *Acta Odontologica Scandinavica* 41: 355–62.
Norén, J. G., B. O. Magnusson, and H. Grahnén. 1978. Mineralisation defects of primary teeth in intra-uterine undernutrition. II. A histological and microradiographic study. *Swedish Dental Journal* 2: 67–72.
Norén, J. G., H. Odelius, B. Rosander, and A. Linde. 1984. SIMS analysis of deciduous enamel from normal full-term infants, low birth weight infants and from infants with congenital hypothyroidism. *Caries Research* 18: 242–49.

Ogrinc, N., and M. Budja. 2005. Paleodietary reconstruction of a neolithic population in Slovenia: A stable isotope approach. *Chemical Geology* 218: 103–16.

Oribasius. 1933. *Collectionum Medicarum Reliquiae*, trans. J. Raeder. Berlin: Teubner Press.

Ortner, D. J., and S. Mays. 1998. Dry-bone manifestations of rickets in infancy and early childhood. *International Journal of Osteoarchaeology* 8 (1): 45–55.

Owsley, D. W., and R. L. Jantz. 1985. Long bone lengths and gestational age distributions of post-contact period Arikara Indian skeletons. *American Journal of Physical Anthropology* 68: 321–28.

Panter-Brick, C. 1998. *Biosocial Perspectives on Children*, vol. 10. New York: Cambridge University Press.

Parkin, T. G. 1992. *Demography and Roman Society.* Baltimore: Johns Hopkins.

Perry, M. A. 2005. Redefining childhood through bioarchaeology: Toward an archaeological and biological understanding of children in antiquity. *Archaeological Papers of the American Anthropological Association* 15: 89–111.

Prowse, T. L., S. R. Saunders, H. P. Schwarcz, P. Garnsey, R. Macchiarelli, and L. Bondioli. 2008. Isotopic and dental evidence for infant and young child feeding practices in an imperial Roman skeletal sample. *American Journal of Physical Anthropology* 137: 294–308.

Prowse, T. L., H. P. Schwarcz, S. R. Saunders, R. Macchiarelli, and L. Bondioli. 2004. Isotopic paleodiet studies of skeletons from the imperial Roman-age cemetery of Isola Sacra, Rome, Italy. *Journal of Archaeological Science* 31: 259–72.

Rawson, B. 2003. *Children and Childhood in Roman Italy.* Oxford: Oxford University Press.

Reid, D. J., and M. C. Dean 2000. Brief communication: The timing of linear hypoplasias on human anterior teeth. *American Journal of Physical Anthropology* 113: 135–39.

Reid, D. J., S. W. Hillson, and M. C. Dean. 2000. Defining chronological growth standards for known fractions of tooth crown height in primate anterior teeth (abstract). *American Journal of Physical Anthropology* 30 (Supplement): 260.

Richards, M. P., S. A. Mays, and B. T. Fuller. 2002. Stable carbon and nitrogen isotope values of bone and teeth reflect weaning age at the medieval Wharram Percy site, Yorkshire, UK. *American Journal of Physical Anthropology* 119: 205–10.

Salomone F., L. Bondioli, M. Dazzi, G. Geusa, G. Pedicelli, A. Sperduti, and R. Macchiarelli. 1997. Variazioni strutturali attraverso l'età dell'osso corticale e trabecolare nelle popolazioni umane del passato: Rilievi morfometrici, radiografici, tomografici ed elaborazione digitale di immagine (abstract). *XII Congresso degli Antropologi Italiani. Storia del Popolamento del Mediterraneo: Aspetti Antropologici, Archeologici e Demografici, Palermo.*

Sarnat, B. G., and I. Schour. 1941. Enamel hypoplasias (chronic enamel aplasia) in relationship to systemic disease: A chronological, morphological and etiological classification. *Journal of the American Dental Association* 28: 1989–2000.

Saunders, S. R. 2008. Subadult skeletons and growth-related studies. In *Biological Anthropology of the Human Skeleton*, 2nd ed., ed. M. A. Katzenberg and S. R. Saunders, 117–48. New York: John Wiley & Sons Inc.

Saunders, S. R., and R. D. Hoppa. 1993. Growth deficit in survivors and non-survivors: Bio-

logical mortality bias in subadult skeletal samples. *American Journal of Physical Anthropology* 36: 127–51.
———. 2003. Growth deficit in survivors and non-survivors: Biological mortality bias in subadult skeletal samples. *Yearbook of Physical Anthropology* 36: 127–51.
Saunders, S. R., R. D. Hoppa, R. Macchiarelli, and L. Bondioli. 2000. Investigating variability in human dental development in the past. *Anthropologie* 38: 101–7.
Saunders, S. R., R. D. Hoppa, and R. Southern. 1993. Diaphyseal growth in a nineteenth century skeletal sample of subadults from St. Thomas' Church, Belleville, Ontario *International Journal of Osteoarchaeology* 3: 265–81.
Schoeninger, M. J. 1985. Trophic level effects on $^{15}N/^{14}N$ and $^{13}C/^{12}C$ ratios in bone-collagen and strontium levels in bone-mineral. *Journal of Human Evolution* 14: 515–25.
Schoeninger, M. J., and M. J. Deniro. 1984. Nitrogen and carbon isotopic composition of bone collagen from marine and terrestrial animals. *Geochimica et Cosmochimica Acta* 48: 625–39.
Schurr, M. R. 1997. Stable nitrogen isotopes as evidence for the age of weaning at the Angel site: A comparison of isotopic and demographic measures of weaning age. *Journal of Archaeological Science* 24: 919–27.
———. 1998. Using stable nitrogen-isotopes to study weaning behavior in past populations. *World Archaeology* 30: 327–42.
Schurr, M. R., and M. L. Powell. 2005. The role of changing childhood diets in the prehistoric evolution of food production: An isotopic assessment. *American Journal of Physical Anthropology* 126: 278–94.
Seow, W. K. 1992. Dental enamel defects in low birthweight children. *Journal of Palaeopathology, Monographic Publications* 2: 321–30.
Simpson, S. W. 1999. Reconstructing patterns of growth disruption from enamel microstructure. In *Human Growth in the Past: Studies From Bones and Teeth*, ed. R. D. Hoppa and C. M. FitzGerald, 241–63. Cambridge: Cambridge University Press.
Soranus of Ephesus. 1956. *Gynecology*, trans. O. Temkin. Baltimore: The Johns Hopkins University Press.
Sperduti, A., L. Bondioli, T. Prowse, F. Salomone, D. Yang, R. D. Hoppa, S. R. Saunders, and R. Macchiarelli. 1997. Reconstructing life conditions of the juvenile population of Portus Romae in 2nd–3rd century AD. *International Symposium Humans from the Past: Advancement in Research and Technology.* Rome, Italy.
Sperduti, A. 1995. I resti scheletrici umani della necropoli di età Romano-imperiale di Isola Sacra (I-III Sec DC): Analisi paleodemografica. PhD dissertation. Rome, Università di Roma 'La Sapienza,' Rome, Italy.
Taji, S., T. Hughes, J. Rogers, and G. Townsend. 2000. Localised enamel hypoplasia of human deciduous canines: Genotype or environment? *Australian Dental Journal* 45: 83–90.
Teivens, A. A., H. Mörnstad, J. G. Norén, and E. Gidlund. 1996. Enamel incremental lines as recorders for disease in infancy and their relation to the diagnosis of SIDS. *Forensic Science International* 81: 175–83.

Toynbee, J. M. C. 1941. Book review: *La Necropoli Del Porto Di Roma Nell'Isola Sacra. Journal of Roman Studies* 31: 207–9.

Turner, B. L., J. L. Edwards, E. A. Quinn, J. D. Kingston, and D. P. Van Gerven. 2007. Age-related variation in isotopic indicators of diet at medieval Kulubnarti, Sudanese Nubia. *International Journal of Osteoarchaeology* 17: 1–25.

Walker, A. F. 1990. The contribution of weaning foods to protein energy malnutrition. *Nutritional Research Reviews* 3: 25–47.

White, C. D., and H. P. Schwarcz. 1994. Temporal trends in stable isotopes for Nubian mummy tissues. *American Journal of Physical Anthropology* 93: 165–87.

Wiedemann, T. 1989. *Adults and Children in the Roman Empire*. London: Routledge.

Wood, C. 2004. An Investigation of the Prevalence of Rickets Among Subadults From the Roman Necropolis of Isola Sacra, Italy. MA thesis. McMaster University, Hamilton, Ontario, Canada.

Wood, J. W., G. R. Milner, H. C. Harpending, and K. M. Weiss. 1992. The osteological paradox: Problems of inferring prehistoric health from skeletal samples. *Current Anthropology* 33: 343–70.

Wright, L. E., and C. J. Yoder. 2003. Recent progress in bioarchaeology: Approaches to the osteological paradox. *Journal of Archaeological Research* 11: 43–70.

Yurdakok, K., T. Yavus, and C. Taylor. 1990. Swaddling and acute respiratory infections. *American Journal of Public Health* 80: 873–75.

• 9 •

Examining Nutritional Aspects of Bone Loss and Fragility across the Life Course in Bioarchaeology

S. C. Agarwal and B. Glencross

Introduction

Osteoporosis is a growing health concern in the aging populations of developed countries, with a recent count of 75 million people affected in Europe, the United States, and Japan (EFFO and NOF 1997). It is estimated that one in three women and one in four men over age fifty will have an osteoporosis-related fracture in her/his remaining lifetime (NOF 2005; Johnell and Kanis 2006). Osteoporosis is a systemic skeletal disease characterized by a reduction in bone mass and a deterioration of the microstructure of bone tissue, with a consequent increase in bone fragility and susceptibility to fracture. Age-related bone loss is found in both sexes, but is accelerated in females with the onset of menopause. However, a woman's risk of developing osteoporosis is greatly mediated by factors that are independent of the menopause-induced drop in estrogen levels. For example, factors such as genetics, physical activity, parity, and lactation also play an important role in bone maintenance. Diet and nutrition have also long been emphasized in the biomedical literature as important in bone growth, maintenance, and fragility. While dietary nutrients such as calcium, protein, and vitamin D are frequently discussed in popular literature as integral to bone mass and loss, the role of these and other nutrients are complex and possibly synergistic with other lifestyle factors. More importantly, the role and impact of nutrients vary over the human life course. Examining patterns of bone maintenance and fragility in past populations offers a unique opportunity to observe bone health in groups that had very different diets combined with differing social and cultural lifestyles than our own. Bioarchaeological studies provide valuable biocultural and evolutionary insight into the etiology of bone loss and fragility in both living and ancient human populations. In this chapter, we review bone biology and the major influences on bone maintenance and fragility, with an emphasis on the cumulative effects acquired during the life course that lead to adult disease states such as osteoporosis and fragility fracture. Further, we outline a life course model for human bone loss and fragility with emphasis on the influential roles of key factors, particularly diet and nutrition, during specific developmental stages. Finally, with examples in bioarchaeology, we examine the role of nutrition in bone maintenance, and the value of the life course model for studies of both bone loss and fragility fracture in human populations.

Bone Biology: How Bone Works

The growth, maintenance, and deterioration of the individual skeleton occur at the tissue level. Bone is a dynamic tissue that has a remarkable ability to grow, maintain, and renew itself through the coordinated activities of bone cells[1] (osteoblasts, osteoclasts, and osteocytes). In the immature skeleton, the process of continued bone growth is called *modeling* where bone is formed and then soon resorbed in different locations (Parfitt 2003). In a typical long bone, an increase in diameter occurs through new bone apposition on the external surface and removal of bone on the internal surface, while an increase in length occurs with new bone replacing cartilage at the ends of the bone. The periodic replacement of mechanically defective or faulty bone tissue for the purpose of maintaining the adult skeleton occurs through a process called *remodeling*, where bone is systematically resorbed and then replaced locally (Parfitt 2003). However, modeling does not only occur in the juvenile skeleton, and remodeling does begin in childhood (Parfitt 2003).

While modeling typically leads to an overall gain in bone mass, remodeling is associated with the maintenance or eventual loss of bone. Bone tissue that serves a metabolic function will be remodeled as an important part of its role in calcium homeostasis and the production of blood cells in the body. In areas where the main function of bone is to resist mechanical loads, the tissue will be remodeled to maintain or repair the damage that has accumulated with time (Currey 2003; Martin 2003; Parfitt 2003). Bone also remodels to heal fractures and in response to intense physiological demands for calcium as during pregnancy and lactation (Frost 2003; Parfitt 2003). Both growth hormones and cytokines play various roles in coordinating bone remodeling (Rosen 2002), but why and how exactly the bone cells work together to model and remodel are not completely understood. The role of the skeleton as a structural and sensitive homeostatic mechanical organ is now recognized as central to our understanding of bone biology (Martin 2003). The skeleton is able to "adapt" through modeling and remodeling to mechanical loading sensed through the living bone cells, in order to tolerate the effects of activity at the tissue level (Martin 2003). This means that the mechanisms of bone biology can record many aspects of life history at the bone tissue level. Significant lifetime changes related to growth, with nutrition and activity, or disease can be seen through the study of morphology, microstructure, and markers of bone turnover recorded in the skeleton (for more on bone remodeling, modeling, and loading see Frost 2003; Pearson and Lieberman 2004). A loss of bone or differences in bone (micro-) morphology within and between human populations, including past populations, can be used to reconstruct the possible influences on bone biology during life.

A number of factors contribute to the mechanical properties of bone and its ultimate strength or fragility. The primary factors can be collectively grouped as either aspects of bone quantity or bone quality (Fig. 9.1). Bone quantity generally refers to

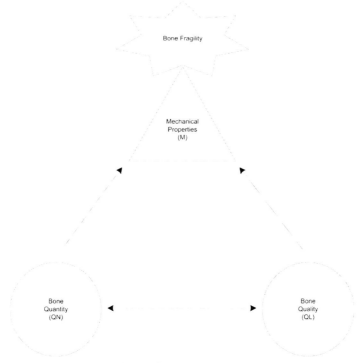

Figure 9.1. Key determinants of bone fragility. Both bone quantity (such as the mass or density) and bone quality (such as bone microstructure and organization, geometric properties, and mineralization) contribute to the mechanical properties of whole bones. The quantity and quality components of a bone are interdependent with changes in either, directly influencing bone strength and ultimately fragility.

the amount of bone, bone density, or mass. While bone mass does contribute to bone fragility, there are a number of additional factors that significantly contribute to bone strength that are independent of bone mass (Heaney 1992; Turner 2002; Watts 2002; Burr 2004). These qualitative aspects of bone include the structure of the bone as a whole, as well as the microstructure and properties of the bony tissue itself. These factors, typically grouped as bone quality, include bone geometry, trabecular and cortical microarchitecture, and bone material properties (such as bone mineralization and microdamage).

Changing Influences on Bone Maintenance and Fragility over the Life Course

Both bone quantity and aspects of quality are strongly affected by the loss of sex hormones (estrogen in women) and aging. However, while bone loss is clearly an age-related and postmenopausal phenomenon, osteoporosis has a multifactorial etiology, and the risk of developing osteoporosis is greatly mediated by factors that are independent of the menopause-induced drop in estrogen levels in females. For example, factors such as heredity, ethnicity, physical activity, parity, and lactation also play an important role in bone maintenance (Stevenson et al. 1989; Sowers and Galuska 1993; Ward et al. 1995; Nelson and Villa 2003; Ralston 2005). Perhaps the most widely discussed and popularly known influence on bone maintenance and fragility is diet and nutrition. Both clinical and anthropological studies (see for example Yano et al. 1985; Hu et al. 1993; Gonzalez-Reimers et al. 2002, 2004, 2007) have emphasized the role of diet and nutrition in bone maintenance. However, current experimental and epidemiological studies (Cooper et al. 2001; Gale et al. 2001; Mehta et al. 2002; Oreffo et al. 2003; Dennison et al. 2005; Ferrari et al. 2006) have also demonstrated that the role and impact of diet and nutrition on bone turnover and fragility varies throughout the life course. For example, the importance and utilization of nutrients in the body, such as calcium, has a different impact on the modeling skeleton of a growing child compared to the fragile skeleton of an elderly individual suffering from diminished calcium absorption in the gut. Current clinical and epidemiological studies on the phenomenon of bone loss and bone deterioration leading to bone fragility have clearly demonstrated the impact of cumulative influences on bone maintenance during life (Barker 1995; Kuh and Ben-Shlomo 1997; Cooper et al. 2005; Pearce et al. 2005). Recent work by social scientists studying chronic diseases and using life course approaches has also emphasized the importance of understanding the interrelationships and joint cumulative contributions of different factors (e.g. genetics, diet, exercise, reproduction) to bone development, maintenance, and loss (Weaver 1998; Agarwal and Stuart-Macadam 2003; Agarwal et al. 2004; Fausto-Sterling 2005).

Bone biology and the mechanical properties of bone tissue vary and change over the life course of an individual. Bone quantity, bone quality, and the kinetics of bone loss leading to bone fragility are determined by the interaction of different genetic and environmental factors experienced across the lifespan (Fig. 9.2). The fac-

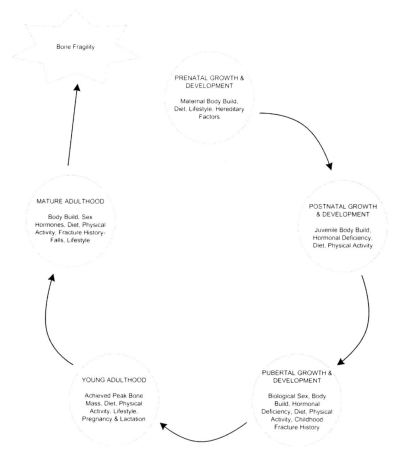

Figure 9.2. Developmental influences on bone fragility from a lifecycle perspective. Adult bone quantity, quality, and the kinetics of bone loss leading to bone fragility are determined by the accumulation of different insults experienced during the growth, development, maintenance and ageing of the skeleton. Diet and nutrition play key roles in bone fragility throughout the entire lifecycle (see text for detailed discussion).

tors implicated in bone loss and fragility such as biological sex, heredity, body build, reproductive history, physical activity, and fracture history play key roles at different points throughout life. More importantly, the influence of each factor at a given life stage is carried forward and layered upon with the unique influences of the subsequent stages. The creation and ultimate nature of skeletal tissues begins in utero and continues to change through the various stages of growth and development. Prenatal growth and development of the skeleton is dependent on maternal body build, diet and nutrition, and hereditary influences. Postnatal growth and development of the skeleton continues to be influenced by the juvenile body build, diet, early growth hormones, and biomechanical strain and activity. Finally, these influences continue to be important during pubertal growth and development, with activity and sex hormones playing particularly key roles during the adolescence stage.

Support for the developmental origins of bone fragility and risk of fracture in later life comes from a variety of sources. Evidence that the prenatal and early postnatal environments can impact the trajectory of skeletal growth is available from experimental rat models and mother/offspring cohort studies in humans. These studies (Namgung and Tsang 2000; Godfrey et al. 2001; Mehta et al. 2002; Oreffo et al. 2003; Harvey et al. 2004; Ganpule et al. 2006) demonstrate that manipulation of the periconceptual, intrauterine, and neonatal environment can lead to altered rates of skeletal growth and persistent structural changes. Most experimental manipulations with animals have involved dietary restriction, specifically low protein. For example, Mehta et al. (2002) found intrauterine exposure to a maternal low-protein diet resulted in reduced bone mass and altered growth plate morphology in late adulthood among the offspring. Oreffo et al. (2003) investigated the effects of maternal protein deficiency on mesenchymal stem cell activity in the developing offspring and found that a maternal low-protein diet delayed the activity of bone-forming cells and subsequent skeletal maturation. In addition, a number of studies have demonstrated that bone mass in human newborns is related to the body build, lifestyle, and nutrition of their pregnant mothers (Namgung and Tsang 2000; Godfrey et al. 2001; Harvey et al. 2004; Ganpule et al. 2006). All of these studies found maternal body build and nutrition influenced fetal nutrient supply and as a consequence bone growth. There are also many epidemiological studies that relate impaired human fetal and early postnatal growth to osteoporosis and an increased risk of fragility fracture in adulthood (Cooper et al. 1997; Cooper et al. 2001; Gale et al. 2001; Dennison et al. 2004; Dennison et al. 2005). The association between weight in infancy and adult bone mass has been repeatedly found in a number of large cohort studies of men and women between the ages of sixty to seventy-five years that show significant relationships between birth weight, weight at one year, and bone mineral content at the lumbar spine and femoral neck as an adult (Cooper et al. 1997; Cooper et al. 2001; Dennison et al. 2005). Further, a study by Cooper et al. (2001) provides direct evidence that low bone mass and increased risk of hip fracture are higher in those adults with a low birth weight and in those who experience slow rates of growth during childhood.

There is also evidence suggesting the period of adolescence to be highly critical for bone acquisition and maintenance. Dietary-related research has found that adolescents suffering from anorexia nervosa with low body mass index, low calcium intake, and low physical activity to be at risk for low bone mineral density in the lumbar spine and femoral neck depending on the duration of their illness and amenorrhea (Castro et al. 2000). Similarly, Parsons et al. (1997) found a macrobiotic diet (vegan-type low in calcium and vitamin D) consumed by adolescents had a negative impact on whole body, spine, proximal femur, and forearm bone mineral content. In a unique study, Ferrari et al. (2006) identified low bone mineral gain during puberty in females with a history of fractures. Their results showed low bone mass at multiple skeletal sites and overall smaller vertebral size at maturity in females with a history of skeletal fracture, which they interpret as "an early marker of persistent bone fragility" (501). While low bone mass in adolescence has been implicated in later bone loss leading to fragility, sex differences in bone acquisition and ultimately bone strength has also recently been highlighted as an important contributor to bone fragility. Sex differences exist where new bone formation on the outer surface of bones persists longer in males, and females experience greater bone formation on the inner surface of bones (Garn 1972). Schoenau et al. (2001) found that during puberty, the longer period of new bone formation in males effectively strengthened their bones while bone mainly laid down on the inner surface by females had little effect on bone stability. Instead, they suggest that bone deposition in females is responsible for creating a "calcium reservoir" to be tapped by mother and infant during pregnancy and lactation without further compromising bone stability during pregnancy.

As the skeleton moves into adulthood, physiological hormones, diet, and physical activity continue to be important in bone maintenance and fragility. However, many of these factors are mediated by physiological and biocultural influences unique to the adult life stages. Individuals that have completed skeletal growth enter the young adult stage with the greatest amount of bone tissue they will achieve during their lifetime, or peak bone mass. It has been shown that greater achieved peak bone mass affords a lower risk of fracture later in life (Matkovic, Badenhop-Stevens, and Crncevic-Orlic 2002). While many factors contribute to achieve peak bone mass, calcium nutrition during adolescent skeletal growth (Matkovic, Badenhop-Stevens, and Crncevic-Orlic 2002) and physical activity during young adulthood (Pearson and Lieberman 2004) seem to have the greatest impact. There are several factors that mediate the amount of bone tissue and its strength in the adult skeleton. For example, additional lifestyle factors such as smoking, alcohol, or medications such as corticosteroids can begin to accelerate bone loss in adults (Stevenson et al 1989; Ward et al. 1995; Rosen 2002). Systemic hormones continue to influence bone maintenance in both adult sexes (Rosen 2002), and the suite of hormones involved in both pregnancy and lactation play a significant role in bone maintenance in young adult females. While pregnancy and lactation are high bone turnover states due to the nutritional demands of the fetus and child, the long-term effect of pregnancy and lactation on bone loss and fragility is not clearly understood. Adaptive changes in calcium absorption in the gut occur during pregnancy, lactation, and weaning, to ensure both adequate calcium supply for fetal growth and maintenance of the maternal skeleton (Specker 2004). Studies of the effects of pregnancy on bone have found conflicting results (Drinkwater and Chesnut 1991; Sowers et al. 1991; Kent et al. 1993; Cross et al. 1995; Naylor et al. 2000); however epidemiological evidence suggests that parity may decrease fracture risk and could increase bone density (Sowers et al. 1992; Fox et al. 1993; Murphy et al. 1994). While longitudinal studies indicate that bone loss can occur during initial lactation (Lamke, Brundin, and Moberg 1977; Chan et al. 1982; Hayslip et al. 1989; Kent et al. 1993; Drinkwater and Chesnut 1991; Sowers et al. 1993; Sowers et al. 1995; Affinito et al. 1996; Lopez et al. 1996; Sowers 1996), there is substantial evidence that recovery of bone occurs with extended lactation and during weaning (Kent et al. 1993; Sowers et al. 1993; Sowers et al. 1995; Affinito et al. 1996; Lopez et al. 1996; Sowers 1996; Pearson et al. 2004).

In mature adulthood and the elderly life stage, the skeleton is a cumulative product of the fetal environment, early growth and development, peak bone mass, and the lifestyle led as a young adult. Both systemic hormones and nutrition continue to be key in both bone maintenance and fracture risk (see also discussion below on the study of nutrition and bone maintenance). Estrogen deficiency is well known as a cause of postmenopausal bone loss in women, and endogenous estrogen along with androgens play a role in male bone loss as well (Rosen 2002; Stini 2003). Several nutrients seem to be important in sustaining bone health in the mature adult particularly protein, calcium, sodium, and phosphorus (Rosen 2002; Stini 2003). Vitamin D has been suggested to be especially critical for maintenance of the elderly skeleton (Rosen 2002, Vieth 2003). Finally, nonskeletal factors can also contribute to fracture risk in old age, notably falls, which are in part also interrelated to muscle strength and previous fracture history and recovery (Rosen 2002).

The pathogenesis of bone loss and fragility fracture is clearly complex and the product of cumulative and synergistic influences over the life course. All the influences work together on the overall health of the individual, the whole skeleton, and at the tissue level through remodeling. Further, interactions at the tissue level and the various environmental factors experienced during the life course are mediated in a broader social and ecological context (Fig. 9.3). Many factors implicated in bone loss and fragility discussed above such as body build, reproductive history, physical activity, and lifestyle are culturally modifiable. Their impact on skeletal growth and development is regulated through personal choices, social and economic relations, and cultural norms that can also vary across the life course. Clinical nutritionists have long recognized the contribution of diet to childhood skeletal growth, as well as the importance of specific dietary requirements in relation to their availability to the individual within the context of the family (Behrman 1995; Davies, Evans, and Gregory 2005). Further, the availability and quality of foods determined by macrolevel factors like community infrastructure and socioeconomic prosperity has also been identified as central to growth of the skeleton in childhood (Matkovic, Badenhop-Stevens, and Crncevic-Orlic 2002), as well as fracture probability in later life (Johnell and Kanis 2006). While bioarchaeologists have examined nutritional influences on bone maintenance and fragility in past populations, influences have rarely been considered in a life course approach. We will briefly discuss some of the key nutrients that have been investigated in bone health in past populations, and then consider the value of utilizing a multilevel framework to examine bone loss and fragility over the life course.

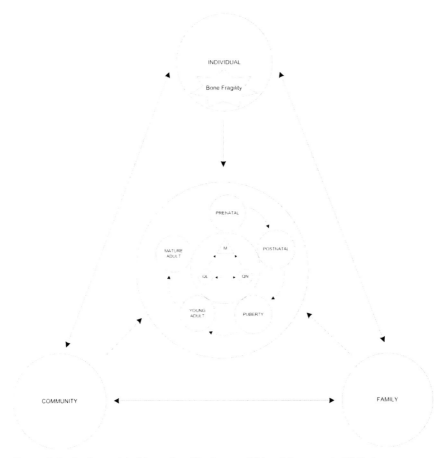

Figure 9.3. Basic model of bone fragility in a multi-level framework. While bone fragility is determined in part by individual biology and direct insults experienced during growth and development, also recognized are social and economic factors influencing bone acquisition and bone loss over the life cycle. Dietary factors affecting bone fragility at the individual, family and community levels might include availability of nutritious foods regulated by interactions between individual choice, and group social and economic relations.

Examples in Bioarchaeology

The Study of Nutrition and Bone Maintenance

While it is recognized that several factors were likely influential in bone maintenance in past populations, a large number of paleopathology studies have focused on the relationship between bone mass and nutrition (see for example, Dewey, Bartley, and

Armelagos 1969a, 1969b; Ericksen 1976; Martin and Armelagos 1979, 1985; Pfeiffer and King 1981; Thompson, Salter, and Laughlin 1981; Thompson et al. 1983; Thompson and Gunness-Hey 1981; Cassidy 1984; Nelson 1984; Velasco-Vazquez et al. 1999; Gonzalez-Reimers et al. 2002, 2004, 2007). Both the amount of dietary calcium and its absorption are thought to have been important factors in bone health in past populations, especially during the transition from hunting and gathering lifestyles to full-scale domestication of plants and animals. While the degree of reliance on domesticated foods varied in different regions, the transition to food production in the Neolithic period (around 12,000–10,000 years ago) in many areas of the world is thought to have been characterized by a reduction in the consumption of animal protein and an increased reliance on foods that often lacked nutritional variety (Nelson, Sauer, and Agarwal 2002; Larsen 2003). Further, overall morbidity and mortality are thought to have increased in farming societies due to the increased prevalence of infectious disease in large sedentary populations living in closer proximity to waste products and animals carrying pathogens, together with poor nutrition and periods of starvation (Nelson, Sauer, and Agarwal 2002). Factors that are thought to have had a negative impact on bone metabolism include a reduction of calcium and iron content in the diet, and the reliance on grains with an increase in phytate, which is thought to bind to and reduce the availability of calcium in the body (Nelson, Sauer, and Agarwal 2002; Larsen 2003). Several paleopathology studies have reported low bone mass in early agricultural populations compared to modern populations (for more extensive reviews see Pfeiffer and Lazenby 1994; Agarwal and Grynpas 1996; Nelson, Sauer, and Agarwal 2002). The earliest studies of bone mass in skeletons from Sudanese Nubia (350–1400 CE) using various methodologies, such as cortical microscopy (Martin 1981; Martin and Armelagos 1979, 1985) and measurement of cortical thickness (Dewey, Bartley, and Armelagos 1969a, 1969b), interpreted the results as reflecting chronic malnutrition. Nutritional hypotheses for bone loss in the past have also been suggested in studies of bone mass in eastern and southwestern North American prehistoric and contact-period skeletal samples. For example, Ericksen (1976) suggested that nutrition was an important determinant of bone loss in her comparative study of Eskimo, Pueblo, and Arikara archaeological samples. She found cortical thinning of the humerus and femur in the Pueblo samples that consumed a primarily cereal-based diet. Studies of bone remodeling in these groups also revealed differences in bone turnover related to dietary differences between the groups, and was interpreted as a consequence of the particularly low-protein diet of the sedentary Pueblo (Richman, Ortner, Schulter-Ellis 1979; Ericksen 1980). A number of subsequent bioarchaeological studies of Native American agricultural populations also noted bone loss suggestive of nutritional stress (Pfeiffer and King 1981; Cassidy 1984; Nelson 1984). More recent studies of low cortical and trabecular bone mass in prehispanic archaeological skeletal remains from Tenerife, Canary Islands, using histomorphometry, quantitative computed tomography (QCT), and x-ray absorptiometry have hypothesized that episodic starvation and low protein and calcium intake are responsible for reduced bone mass (Velasco-Vazquez et al. 1999; Gonzalez-Reimers et al. 2002, 2004, 2007).

Protein is hypothesized to have had a detrimental effect on bone mass and maintenance in the past, specifically the consumption of large amounts of animal protein. High-protein diets can lead to an abnormal increase in the acidity of body fluids causing skeletal calcium to be resorbed in the body as a means to buffer acid loads imposed by the animal proteins (Orwoll 1992), and/or calcium to be bound by the kidneys through sulphates and phosphorus produced in protein metabolism (Schuette et al. 1981). Bone core studies on different archaeological Inuit skeletal samples with traditionally high-protein diets found increased intracortical porosity and bone loss, when compared to modern US Caucasians (Thompson, Salter, and Laughlin 1981; Thompson et al. 1983; Thompson and Gunness-Hey 1981). However, the influence of protein intake on bone metabolism is complex. The increase in urinary calcium excretion due to high dietary protein does utilize the skeleton to neutralize acid generated from a high-protein diet (Kerstetter, O'Brien, and Insogna 2003), and cross-sectional studies of modern populations have shown a positive association between high animal protein intake and hip fracture prevalence (Abelow, Holford, and Insogna 1992). In contrast, other studies indicate that dietary protein can have a positive effect on bone metabolism. Dietary protein can increase circulating levels of insulin-like growth factor-1 (IGF-1), affecting skeletal development through increasing bone formation (Dawson-Hughes 2003; Bonjour 2005). Dawson-Hughes has suggested that the effect of protein consumption on bone may depend on its potential interaction with calcium intake. A higher intake of calcium would simply give more absorbed calcium to offset urine losses, and by further lowering bone turnover, could also reduce the effects of mild acidosis or amplify the positive effects of IGF-1 on bone maintenance (2003). The effects of dietary protein on bone seem to be dose dependent, and low intake could be deleterious particularly among susceptible groups such as the elderly. High intakes are known to be equally harmful, and depend on calcium intake and possibly also vitamin D intake to mediate the impact of dietary protein on bone (Dawson-Hughes 2003; Kerstetter, O'Brien, and Insogna 2003).

Calcium intake also needs to be considered along with the roles of other nutrients. The influence of phosphorus on calcium absorption and its role in bone maintenance has been considered in the study of early agriculturalists. Cultivated cereal-based diets may have had a high ratio of phosphorus to calcium, which has been hypothesized to inhibit calcium absorption (Nelson 1984; Cohen 1989). Both calcium and phosphorus have a relationship with parathyroid hormone (PTH), and studies have suggested that increased phosphorus intake could adversely affect bone metabolism (with increased bone resorption and less bone formation) through increased PTH secretion when specifically combined with low calcium intake (Kemi, Karkkainen, and Lamberg-Allardt 2006; Palacios 2006). However, it appears the ratio of phosphorus to calcium intake is key rather than the absolute levels of phosphorus intake alone (Palacios 2006), and clinical studies have not demonstrated a link between high phosphorus intake and bone loss in healthy humans with normal calcium intakes (Calvo 1994; Palacios 2006). The relationship between calcium intake and bone maintenance is also complex, and calcium absorption in the gut and excretion can potentially contribute more to calcium balance than intake alone (Nordin et al. 1995).

Perhaps a more influential nutrient on bone metabolism in the past was vitamin D. Some have suggested that osteoporosis may be a consequence of subclinical mild vitamin D deficiency (Parfitt 1990; Heaney et al. 1999; Vieth 2003). Vitamin D deficiency has been considered in interpretations of bone loss in historical skeletal samples (Mays 1996; Agarwal et al. 2004), but the effect of mild vitamin D insufficiency on bone maintenance, perhaps more common in some historical populations, has not been widely examined. Human diets have varied throughout history, and it is clear from this discussion that the influence of calcium, protein, vitamin D, and other nutrients on skeletal health is complex. Future nutritional hypotheses of bone loss in the past must consider the current biomedical research and the synergistic effects of other biological environmental influences on bone maintenance.

Bone Maintenance over the Life Course

Both past and living human populations show varying patterns in bone maintenance and the degree of bone loss in old age. While diet and nutrition are likely to have played a significant role in skeletal health in prehistoric and historic times, bone maintenance is influenced by multiple factors including growth, activity, and reproduction. While the emphasis is on osteoporotic bone loss in old age, we are beginning to realize that our understanding of long-term fragility, in both past and present populations, can only be achieved through the understanding of bone maintenance throughout the life course (Weaver 1998; Agarwal and Stuart Macadam 2003; Fausto-Sterling 2005). There are two major events in the life course intricately linked to nutrition that are known to affect bone loss and fragility in later life: the achievement of peak bone mass and reproduction. The accumulation of bone mass and its modeling during growth and development are directly related to the achievement of peak bone mass, and in turn the total amount of bone available to remodel and lose in later life. In females, reproductive factors, specifically parity and lactation, are known to play significant roles in bone maintenance and loss in young adulthood, as well as postmenopausal bone loss and fragility.

While there have been many studies of bone loss and fragility in old age in archaeological samples, there has been little examination of the effect of growth, modeling, and reproduction on the mature skeleton. The classic studies of bone loss in prehistoric Sudanese Nubia were some of the first studies to consider and compare bone growth and maintenance in both juvenile and adult skeletons. Armelagos et al. (1972) found femoral cortical bone loss to be more significant in Nubian females than males, and bone loss to occur at an earlier age when compared to modern populations. They suggest this pattern is related to females achieving lower peak bone mass than males and likely due to early growth disturbance. Armelagos et al. (1972) further suggest that the bone loss observed in young adult females may be due to nutritional (calcium) deficiency and prolonged lactation. Interestingly, while the authors also found an age-related decrease in femoral trabeculae in both sexes, females showed an increase in the thickness of some of the supporting trabecular elements, suggestive of a compensatory mechanism for the loss of trabeculae (Armelagos et al. 1972). A study of cortical bone growth maintenance in prehistoric juvenile Nubians from the Kulubnarti site found that while bone mineral content increases after birth, processes of modeling combined with likely periods of nutritional stress caused a reduction in percent cortical area during early and late childhood (Ven Gerven, Hummert, and Burr 1985). However, the authors also found the quality of bone unchanged during growth, with the cross-sectional geometric integrity (and bending strength) remaining intact through childhood. Unfortunately, this study did not comment on the role of early bone maintenance on later femoral bone loss. Kneissel et al. (1997) conducted a novel study of vertebral cancellous bone growth and loss in a juvenile and adult skeletal sample from medieval Nubia, finding the largest bone trabecular volume present during adolescence when the rod-like trabeculae of childhood begin to change to plate-like structures. Age-related loss of trabecular structure was observed in adults, with changes occurring earlier in life than seen in modern populations.

Two recent studies of European skeletal samples examined bone growth and stress during childhood in comparison to adult bone maintenance. McEwan, Mays, and Blake (2005) examined the correlation of radial bone mineral density (BMD) to indicators of growth disturbances (specifically cortical index (CI), Harris lines, and cribra orbitalia) typically attributed to poor nutrition, in juvenile skeletons from a medieval British sample. McEwan, Mays, and Blake (2005) found BMD to be well correlated with overall growth but not strongly correlated with CI, the latter showing greater sensitivity to environmental stress, as seen in the prehistoric Nubian study by Ven Gerven, Hummert, and Burr (1985), discussed above. McEwan, Mays, and Blake (2005) suggest that bone mineral accrual is relatively stable despite environment (nutritional) stress as compared to cortical bone modeling, and a similar finding was also noted in the femur cortical index measurement (Mays 1999). These studies suggest that long bone cortical thickness is strongly influenced by nutritional stress and that reduced bone quantity during modeling can be carried into adulthood exacerbating later loss of bone density. An additional study of cortical bone loss in adult skeletons from Wharram Percy also showed age-related bone loss (Mays 1999). A recent study by Rewekant (2001) examined the correlation of adult cortical bone loss with the occurrence of indicators of growth disturbances (specifically compression of the skull base and vertebral stenosis) in two Polish medieval skeletal samples with differing socioeconomic status. Rewekant (2001) found greater adult age-related cortical bone loss in the metacarpals from the sample that also showed greater disturbance of bone growth during childhood. Interestingly, there was also less sexual dimorphism in measurements of metacarpal cortical bone and skull base height in the sample that appeared to have suffered greater environmental stress during growth (Rewekant 2001). This study again confirms a relationship between the accumulation of bone during growth, disturbance in the achievement of peak bone mass in both sexes, and loss of bone quantity in later life.

Pregnancy and lactation are also major life course events that have been considered important in interpreting bone maintenance and loss in the past. Early studies of bone loss in the archaeological skeletons of Nubian females were thought to reflect pregnancy and lactation stress (Martin and Armelagos 1979, 1985; Martin et al. 1984; Martin, Goodman, and Armelagos 1985). More recent studies of low bone-mineral density in young female medieval skeletons from Denmark (Poulsen et al. 2001) and Norway (Turner-Walker, Syversen, and Mays 2001; Mays, Brickley, and Ives 2006) have hypothesized that this is a result of insufficient nutrition together with pregnancy and lactation stress. As discussed earlier, however, while pregnancy and lactation are high bone turnover states, the long-term effect of pregnancy and lactation on maternal bone maintenance and fragility is not clearly understood. In contrast to the studies of bone mineral density, Vogel et al.'s (1990) study of a historical European skeletal sample found that female skeletons showed better trabecular connectivity when compared to modern populations, which was attributed to the benefits of high parity. Similarly, Agarwal et al. (2004) have suggested that the unusual patterns of trabecular bone loss seen at Wharram Percy and other archaeological sites, such as little loss observed between middle and old age, lack of significant sex difference in loss, and few fragility fractures, may be related to the long-term, perhaps protective, effects of high parity and extended lactation. Historic practices of high parity and prolonged lactation would have created a very different hormonal milieu for women in past populations as compared to modern Western women. While historic women of reproductive age, such as those found at Wharram Percy, may have suffered from transient bone loss, these losses may not have affected long-term postmenopausal bone fragility (Agarwal and Stuart-Macadam 2003; Agarwal et al. 2004). Further, the less dramatic postmenopausal drop in bone mass and density seen in many archaeological samples may reflect the overall lower lifetime levels of steroid exposure in historical women due to repeated pregnancies and prolonged periods of breastfeeding. The pattern of bone loss seen in modern women may be related to the sudden down regulation of bone-forming osteoblast cells that are elevated with chronically high levels of estrogen and other hormones (Weaver 1998; Agarwal et al. 2004). It seems likely that reproductive factors earlier in the life course would have played a role in bone maintenance in prehistoric and historic females that would have had a significant impact on bone loss and fragility in the later stages of life. Bioarchaeological studies are limited to observing bone loss in isolated individuals or cross-sections of populations. However, while longitudinal study of bone loss over the life course in the past is not possible, age- and sex-related changes in bone quantity and quality can be considered over the life course in past populations (Fig.9.4).

Nutritional Aspects of Bone Loss and Fragility in Bioarchaeology • 211

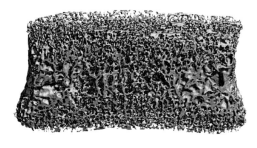
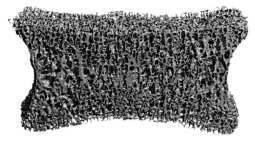
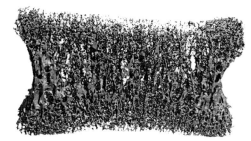

Figure 9.4. High resolution pQCT images of age-related changes in trabecular architecture in fourth lumbar vertebrae from Neolithic archaeological skeletons from Çatalhöyük, Turkey. 3D surface rendered images of the segmented trabecular bone of multiple coronal slices at full resolution (3072x3072, 41μm pixel). Top, juvenile individual showing high number of uniform platelike trabeculae; middle, adult individual showing the typical zonal and anisotropic organization with densely packed trabeculae horizontally and vertically oriented in the superior and inferior region; bottom, mature adult individual showing significant age-related thinning and loss of platelike trabecular elements. Images courtesy of Sabrina Agarwal.

It is increasingly evident that our understanding of bone maintenance in the past can only be achieved with a consideration of factors such as diet and nutrition over the entire life course.

Fragility Fractures and Nutrition

Fragility fractures are the result of underlying disease that adversely affects bone maintenance, quantity, and/or quality weakening the skeleton and making it more vulnerable to injury. Nutritional deficiencies and lifestyle factors that cause disturbances in bone metabolism are extremely important to our understanding of fracture patterns in both modern and ancient peoples. For example, vitamin C deficiency has long been regarded clinically as the cause of the metabolic disease, scurvy and vitamin D deficiency the cause of metabolic disorders, rickets and osteomalacia (Stuart-Macadam

1989). A deficiency of either nutrient can have deleterious effects on bone formation leading to mechanical instability and susceptibility to fracture. Not surprisingly, bioarchaeological studies have frequently recognized the impact of diet and lifestyle contributing to these specific metabolic disorders in past peoples (see for example Stuart-Macadam 1988, 1989; Ortner, Kimmerle, and Diez 1999; Ortner et al. 2001; Brickley and Ives 2006; Mays 2006; Brickley, Mays, and Ives 2007).

More recently, fragility fractures resulting from the metabolic disorder osteoporosis as witnessed in archaeological skeletal samples has received considerable attention (Brickley 2002, 2006; Mays 2006; Mays, Brickley, and Ives 2006). This interest arises from the need to better understand the natural history of osteoporosis and osteoporosis-related fracture, now major health concerns in North America, Europe, and Japan (EFFO and NOF 1997). In modern populations osteoporosis-related fractures typically occur in but are not limited to the wrist (distal radius), spine (vertebra), and hip (proximal femur) (Brown and Josse 2002). Kanis et al. (2001) has shown that the risk of fracture at each of these skeletal sites increases dramatically with age in adults between fifty and eighty years old. The exception is the wrist in males, where fracture risk remains relatively stable through the later years. Further, the risk of osteoporotic fracture at any skeletal site is always greater in females than males. The single most common cause of osteoporotic fracture is a fall from standing height easily producing significant morbidity and mortality in modern populations and a significant reduction in quality of life amongst the elderly (Brown and Josse 2002).

To date, work by bioarchaeologists has mainly concentrated on the paleoepidemiology of osteoporosis-related fractures (Kelley 1980; Lovejoy and Heiple 1981; Glencross 2003; Brickley 2006; Mays 2006). The examination of intraskeletal, age and sex-related fracture patterns has caused some bioarchaeologists (Lovejoy and Heiple 1981; Agarwal and Grynpas 1996; Brickley 2002) to suggest a paucity of osteoporosis-related fractures in the past when compared to the pattern established for modern populations. Various reasons have been suggested for the absence of osteoporosis-related fracture in skeletal samples, such as inherent difficulties in recognizing examples in archaeological remains, and in the past a smaller proportion of individuals living beyond age seventy when senile osteoporosis and hip fracture risk is greatest. These explanations are mainly concerned with diagnostic error and/or selection bias introduced through the nature and demographic composition of skeletal samples.

Perhaps this is why still fewer bioarchaeological studies have gone on to explore the complex etiology of osteoporotic fractures although those that have suggest nutritional hypotheses. For example, in an early paper by Snow (1948), he suggests poor nutrition as an explanation for the pattern of osteoporotic bone lesions observed in the adults and younger individuals of the archaic Indian Knoll skeletal sample. A later study by Kelley (1980) of the same complex hunter-gatherers of Indian Knoll also suggests nutrition as the underlying cause of the osteoporosis-related bone lesions and fractures observed, specifically suggesting diet and water resources lacking in calcium as the root cause. In another unique study, Sievänen et al. (2007) examines and compares the macroanatomy of the proximal femur found in medieval skeletons to that found in contemporary skeletons with and without osteoporosis-related hip fracture. They suggest that changes observed in femoral neck length and circumference reflect temporal trends towards improved general nutrition during growth and a concomitant decline in physical activity predisposing modern individuals to hip fractures caused by a fall from standing height.

The pathogenesis of osteoporosis-related bone loss and fragility fracture is clearly complex and the product of cumulative and synergistic influences over the life course. Nutritional influences remain prominent, with vitamin D deficiency a key component. As discussed earlier, Vieth (2003) notes that mild vitamin D deficiency that is chronic or recurrent over the life course is likely an important factor contributing to osteoporosis-related fracture in later life. Also important is the link he suggests between an array of vitamin D deficiencies known to occur throughout life and their effects on skeletal maintenance and fragility. Bioarchaeological studies of osteoporosis must consider not only fragility fractures observed in elderly females but also the bony lesions and fractures indicative of the various expressions of vitamin D deficiency in both sexes and all ages to better understand the phenomenon of osteoporosis in the past.

However, still to be addressed in bioarchaeological studies is whether fragility fractures in the past occurred as frequently as in modern populations and how this relates to synergistic and cumulative influences on bone metabolism over the life course. To further strengthen our understanding of bone loss, fragility, and fracture in the past, bioarchaeologists will benefit from adopting a life course approach that is already well established in experimental and epidemiological research (see Kuh and Ben-Shlomo 1997; Fausto-Sterling 2005). Recent research in both these areas have shown the developmental origins of bone fragility and the impact of cumulative effects experienced across the life cycle. A better understanding of the epidemiology of osteoporotic fracture in the past and the present can be gained through the examination of fracture patterns in conjunction with data on bone mass, architecture, lifestyle, and nutrition in both sexes and including all age grades.

Conclusion and Future Directions

Bone quantity and quality of the human skeleton is the cumulative product of diet and nutrition over the life course. Further, while nutrition plays a key role in skeletal growth and maintenance during all developmental stages, diet and nutrition are intricately woven with other biological, social, and cultural influences on the skeleton. Bone tissue is a dynamic medium that is capable of recording many life history events, allowing us to reconstruct some of the key influences on bone loss and fragility in living and past populations. While the influence of nutrition on bone quantity has been examined in many archaeological samples, the study of both bone quantity and bone quality have only rarely been examined together with patterns of true fragility-related fractures. Current experimental and epidemiological research clearly highlights the importance of the developmental origins of bone fragility and the importance of cumulative effects experienced across the life course. We propose that a more holistic picture of bone maintenance in the past can be gained with the joint examination of both patterns of bone strength and fragility fracture, and the use of evidence gathered from multiple lines of inquiry, in both sexes and including all age grades. While challenging, the examination of bone maintenance and fragility over the life course in past populations can offer us a fresh look at the role of nutrition in bone health.

Notes

1. The three types of cells responsible for growth and maintenance of bone tissue are: osteoblasts or bone-forming cells; osteoclasts, considered agents of bone resorption or removal; and osteocytes, the mature, living bone cells found embedded in bone tissue.

References

Abelow, B. J., T. R. Holford, and K. L. Insogna. 1992. Cross-cultural association between dietary animal protein and hip fracture: A hypothesis. *Calcified Tissue International* 50 (1): 14–18.

Affinito, P., G. A. Tommaselli, C. di Carlo, F. Guida et al. 1996. Changes in bone mineral density and calcium metabolism in breastfeeding women: A one year follow-up study. *Journal of Clinical Endocrinology and Metabolism* 81: 2314–18.

Agarwal, S. C., M. Dumitriu, G. A. Tomlinson, and M. D. Grynpas. 2004. Medieval trabecular bone architecture: The influence of age, sex, and lifestyle. *American Journal of Physical Anthropology* 124 (1): 33–44.

Agarwal, S. C., and M. D. Grynpas. 1996. Bone quantity and quality in past populations. *The Anatomical Record* 246: 423–32.

Agarwal, S. C., and P. Stuart-Macadam. 2003. An evolutionary and biocultural approach to understanding the effects of reproductive factors on the female skeleton. In *Bone Loss and Osteoporosis: An Anthropological Perspective*, ed. S. C. Agarwal and S. D. Stout, 105–16. New York: Kluwer Plenum Academic Press.

Armelagos, G. J., J. H. Mielke, K. H. Owen, D. P. Van Gerven et al. 1972. Bone growth and development in prehistoric populations from Sudanese Nubia. *Journal of Human Evolution* 1: 89–119.

Barker, D. J. P. 1995. The fetal origins of adult disease. *Biological Sciences* 262: 37–43.

Behrman, J. R. 1995. Household behavior, preschool child health and nutrition, and the role of information. In *Child Growth and Nutrition in Developing Countries*, ed. P. Pinstrup-Andersen, D. Pelletier, and H. Alderman, 32–52. Ithaca, NY: Cornell University Press.

Bonjour, J. P. 2005. Dietary protein: An essential nutrient for bone health. *Journal of the American College of Nutrition* 24 (6 Suppl): 526S–536S.

Brickley, M. 2002. An investigation of historical and archaeological evidence for age-related bone loss and osteoporosis. *International Journal of Osteoarchaeology* 12: 364–71.

———. 2006. Rib fractures in the archaeological record: A useful source of sociocultural information? *International Journal of Osteoarchaeology* 16: 61–75.

Brickley, M., and R. Ives. 2006. Skeletal manifestations of infantile scurvy. *American Journal of Physical Anthropology* 129: 163–72.

Brickley, M., S. Mays, and R. Ives. 2007. An investigation of skeletal indicators of vitamin D deficiency in adults: Effective markers for interpreting past living conditions and pollution levels in 18th and 19th century Birmingham, England. *American Journal of Physical Anthropology* 132: 67–79.

Brown, J. P., and R. G. Josse. 2002. Clinical practice guidelines for the diagnosis and management of osteoporosis in Canada. *Canadian Medical Association Journal* 167: S1–34.

Burr, D. B. 2004. Bone quality: Understanding what matters. *Journal of Musculoskeletal Neuronal Interactions* 4 (2): 184–86.

Calvo, M. S. 1994. The effects of high phosphorus intake on calcium homeostasis. In *Advances in Nutritional Research,* ed. H. H. Draper, 183–207. New York: Plenum Press.

Cassidy, C. M. 1984. Skeletal evidence for prehistoric subsistence adaptation in the central Ohio River Valley. In *Paleopathology at the Origins of Agriculture,* ed. M. N. Cohen and G. J. Armelegos, 307–45. New York: Academic Press.

Castro, J., L. Lázaro, F. Pons, I. Halperin et al. 2000. Predictors of bone mineral density reduction in adolescents with anorexia nervosa. *Journal of the American Academy of Child and Adolescent Psychiatry* 39: 1365–70.

Chan, G. M., P. Slater, N. Ronald, C. C. Roberts et al. 1982. Bone mineral status of lactating mothers of different ages. *Obstetrics and Gynecology* 144: 438–41.

Cohen, M. N. 1989. *Health and the Rise of Civilization.* New Haven, CT: Yale University Press.

Cooper, C., J. G. Eriksson, T. Forsén, C. Osmond et al. 2001. Maternal height, childhood growth and risk of hip fracture in later life: A longitudinal study. *Osteoporosis International* 12: 623–29.

Cooper, C., C. Fall, P. Egger, R, Hobb et al. 1997. Growth in infancy and bone mass in later life. *Annals of Rheumatic Diseases* 56: 17–21.

Cooper, C., K. Javaid, S. Westlake, N. Harvey, and E. Dennison. 2005. Developmental origins of osteoporotic fracture: The role of maternal vitamin D insufficiency. *Journal of Nutrition* 135: 2728S–2734S.

Cooper, C., K. Walker-Bone, N. Arden, and E. Dennison. 2000. Novel insights into the pathogenesis of osteoporosis: The role of intrauterine programming. *Rheumatology* 39: 1312–15.

Cross, N. A., L. S. Hillman, S. H. Allen, G. F. Krause et al. 1995. Calcium homeostasis and bone metabolism during pregnancy, lactation, and postweaning: A longitudinal study. *American Journal of Clinical Nutrition* 61: 514–23.

Currey, J. D. 2003. The many adaptations of bone. *Journal of Biomechanics* 36 (10): 1487–95.

Davies, J. H, B. A. J. Evans, and J. W. Gregory. 2005. Bone mass acquisition in healthy children. *Archives of Disease in Childhood* 90: 373–78.

Dawson-Hughes, B. 2003. Calcium and protein in bone health. *Proceeding of the Nutrition Society* 62 (2): 505–9.

Dennison, E. M., H. E Syddall, S. Rodriguez, A. Voropanov et al. 2004. Polymorphism in the growth hormone gene, weight in infancy, and adult bone mass. *Journal of Clinical Endocrinology and Metabolism* 89: 4898–903.

Dennison, E. M., H. E. Syddall, A. A. Sayer, H. J. Gildbody et al. 2005. Birth weight and weight at one year are independent determinants of bone mass in the seventh decade: The Hertfordshire cohort study. *Pediatric Research* 57: 582–86.

Dewey, J., M. Bartley, and G. Armelagos. 1969a. Femoral cortical involution in three Nubian archaeological populations. *Human Biology* 41: 13–28.

Dewey, J., M. Bartley, and G. Armelagos. 1969b. Rates of femoral cortical bone loss in two Nubian populations: Utilizing normalized and non-normalized data. *Clinical Orthopaedics and Related Research* 65: 61–66.

Drinkwater, B. L., and C. H. Chesnut. 1991. Bone density changes during pregnancy and lactation in active women: A longitudinal study. *Journal of Bone and Mineral Research* 14: 153–60.

EFFO and NOF. 1997. Who are candidates for prevention and treatment for osteoporosis? *Osteoporosis International* 7: 1.

Ericksen, M. F. 1976. Cortical bone loss with age in three Native American populations. *American Journal of Physical Anthropology* 45: 443–52.

Ericksen, M. F. 1980. Patterns of microscopic bone remodeling in three aboriginal American populations. In *Early Native Americans: Prehistoric Demography, Economy, and Technology*, ed. D. L. Brownman, 239–70. The Hague: Houton.

Fausto-Sterling, A. 2005. The bare bones of sex: Part 1- sex and gender. *Signs Journal of Women in Culture and Society* 30: 1491–527.

Ferrari, A., T. Chevalley, J. Bonjour, and R. Rizzoli. 2006. Childhood fractures are associated with decreased bone mass gain during puberty: An early marker of persistent bone fragility? *Journal of Bone and Mineral Research* 21: 501–7.

Fox, K. M., J. Magaziner, R. Sherwin, J. C. Scott et al. 1993. Reproductive correlates of bone mass in elderly women: Study of osteoporotic fractures research group. *Journal of Bone and Mineral Research* 8: 901–8.

Frost, H. 2003. On changing views about age-related bone loss. *Bone Loss and Osteoporosis: An Anthropological Perspective*, ed. S. C. Agarwal and S. D. Stout, 19–30. New York: Kluwer Plenum Academic Press.

Gale, C. R., C. N. Martyn, S. Kellingray, R. Eastell, and C. Cooper. 2001. Intrauterine programming of adult body composition. *Journal of Clinical Endocrinology and Metabolism* 86: 267–72.

Ganpule, A., C. S. Yajnik, C. H. D. Fall, S. Rao et al. 2006. Bone mass in Indian children—relationships to maternal nutritional status and diet during pregnancy: The Pune maternal nutrition study. *Journal of Clinical Endocrinology and Metabolism* 91: 2994–3001.

Garn, S. M. 1972. The course of bone gain and phases of bone loss. *Orthopedic Clinics of North America* 3: 503–20.

Glencross, B. 2003. "An Approach to the Epidemiology of Bone Fractures: Methods and Techniques Applied to Long Bones from the Indian Knoll Skeletal Sample, Kentucky." PhD dissertation, University of Toronto.

Godfrey, K., K. Walker-Bone, S. Robinson, P. Taylor et al. 2001. Neonatal bone mass: Influence of parental birthweight, maternal smoking, body composition and activity during pregnancy. *Journal of Bone and Mineral Research* 16: 1694–703.

Gonzalez-Reimers, E., M. A. Mas-Pascual, M. Arnay-de-la-Rosa, J. Velasco-Vazquez et al. 2004. Noninvasive estimation of bone mass in ancient vertebrae. *American Journal of Physical Anthropology* 125 (2): 121–31.

Gonzalez-Reimers, E., J. Velasco-Vázquez, M. Arnay-de-la-Rosa, and M. Machado-Calvo. 2007. Quantitative computerized tomography for the diagnosis of osteopenia in prehistoric skeletal remains. *Journal of Archaeological Science* 34 (4): 554–61.

Gonzalez-Reimers, E., J. Velasco-Vazquez, M. Arnay-de-la-Rosa, F. Santolaria-Fernandez et al. 2002. Double-energy x-ray absorptiometry in the diagnosis of osteopenia in ancient skeletal remains. *American Journal of Physical Anthropology* 118 (2): 134–45.

Harvey, N. C. W., M. K. Javaid, P. Taylor, S. R. Crozier et al. 2004. Umbilical cord calcium and maternal vitamin D status predict different lumbar spine bone parameters in the offspring at 9 years. (abstract) *Journal of Bone and Mineral Research* 19: 1032.

Hayslip, C. C., T. A. Klein, H. L. Wray, and W. E. Duncan. 1989. The effects of lactation on bone mineral content in healthy postpartum women. *American Journal of Obstetrics and Gynecology* 73: 588–92.

Heaney, R. P. 1992. The natural history of vertebral osteoporosis. Is low bone mass an epiphenomenon? *Bone* 13: S23–S26.

Heaney, R. P., D. A. McCarron, B. Dawson-Hughes, S. Oparil et al. 1999. Dietary changes favorably affect bone remodeling in older adults. *Journal of the American Dietetic Association* 99 (10): 1228–33.

Hu, J., X. Zhao, J. Jia, B. Parpia et al. 1993. Dietary calcium and bone density among middle-aged and elderly women in China. *American Journal of Clinical Nutrition* 58: 219–27.

Johnell, O., and J. A. Kanis. 2006. An estimate of the worldwide prevalence and disability associated with osteoporotic fractures. *Osteoporosis International* 17 (12): 1726–33.

Kanis, J. A., O. Johnell, A. Oden, A. Dawson et al. 2001. Ten year probabilities of osteoporotic fractures according to BMD and diagnostic thresholds. *Osteoporosis International* 12: 989–95.

Kelley, M. A. 1980. Disease and Environment: A Comparative Analysis of Three Early American Indian Skeletal Collections. PhD dissertation, Case Western Reserve University.

Kemi, V. E., M. U. Karkkainen, and C. J. Lamberg-Allardt. 2006. High phosphorus intakes acutely and negatively affect Ca and bone metabolism in a dose-dependent manner in healthy young females. *British Journal of Nutrition* 96 (3): 545–52.

Kent, G. N., R. I. Price, D. H. Gutteridge, J. R. Allen et al. 1993. Effect of pregnancy and lactation on maternal bone mass and calcium metabolism. *Osteoporosis International* S1: 44–47.

Kerstetter, J. E., K. O. O'Brien, and K. L. Insogna. 2003. Dietary protein, calcium metabolism, and skeletal homeostasis revisited. *American Journal of Clinical Nutrition* 78 (S3): 584S–592S.

Kneissel, M., P. Roschger, W. Steiner, D. Schamall et al. 1997. Cancellous bone structure in the growing and aging lumbar spine in a historic Nubian population. *Calcified Tissue International* 61: 95–100.

Kuh, D., and Y. Ben-Shlomo. 1997. *A Life Course Approach to Chronic Disease Epidemiology*. Oxford: Oxford Medical Publications.

Lamke, B., J. Brundin, and P. Moberg. 1977. Changes of bone mineral content during pregnancy and lactation. *Acta Obstetricia et Gynecologica* 56: 217–19.

Larsen, C. S. 2003. Animal source foods and human health during evolution. *Journal of Nutrition* 11 (S2): 3893S–3897S.

Lopez, J. M., G. Gonzalez, V. Reyes, C. Campino et al. 1996. Bone turnover and density in healthy women during breastfeeding and after weaning. *Osteoporosis International* 6: 153–59.

Lovejoy, C. O., and K. G. Heiple. 1981. The analysis of fractures in skeletal populations with an example from the Libben Site, Ottowa County, Ohio. *American Journal of Physical Anthropology* 55: 529–41.

Martin, D. L. 1981. Microstructural examination: Possibilities for skeletal analysis. In *Biocultural Adaptation: Comprehensive Approaches to Skeletal Analysis*, ed. D. L. Martin and M. P. Bumstead, 96–107. Research Reports No. 20. Amherst: Department of Anthropology, University of Massachusetts.

Martin, D. L., and G. J. Armelagos. 1979. Morphometrics of compact bone: An example from Sudanese Nubia. *American Journal of Physical Anthropology* 51: 571–78.

Martin, D. L., and G. J. Armelagos. 1985. Skeletal remodeling and mineralization as indicators of health: An example from prehistoric Sudanese Nubia. *Journal of Human Evolution* 14: 527–37.

Martin, D. L., G. J. Armelagos, A. H. Goodman, and D. P. van Gerven. 1984. The effects of socioeconomic change in prehistoric Africa: Sudanese Nubia as a case study. In *Paleopathology at the Origins of Agriculture*, ed. M. N. Cohen and G. J. Armelagos, 193–214. New York: Academic Press.

Martin, D. L., A. H. Goodman, and G. J. Armelagos. 1985. Skeletal pathologies as indicators of quality and quantity of diet. In *The Analysis of Prehistoric Diets*, ed. R. I. Gilbert and J. H. Mielke, 227–79. New York: Academic Press.

Martin, R. B. 2003. Functional adaptation and fragility of the skeleton. In *Bone Loss and Osteoporosis: An Anthropological Perspective*, ed. S. C. Agarwal and S. D. Stout, 121–36. New York: Kluwer Plenum Academic Press.

Matkovic, V., N. Badenhop-Stevens, E. Ha, Z. Crncevic-Orlic et al. 2002. Nutrition and bone health in children and adolescents. *Clinical Review in Bone and Mineral Metabolism* 1 (3–4): 233–48.

Mays, S., M. Brickley, and R. Ives. 2006. Skeletal manifestations of rickets in infants and young children in a historic population from England. *American Journal of Physical Anthropology* 129: 362–74.

Mays, S. A. 1996. Age-dependent cortical bone loss in a medieval population. *International Journal of Osteoarchaeology* 6: 144–54.

Mays, S. A. 1999. Linear and appositional long bone growth in earlier human populations. In *Human Growth in the Past: Studies from Bones and Teeth,* ed. R. D. Hoppa and C. M. FitzGerald, 290–312. New York: Cambridge University Press.

Mays, S. A. 2006. A palaeopathological study of Colles' fracture. *International Journal of Osteoarchaeology* 16: 415–28.

McEwan, J. M., S. Mays, and G. M. Blake. 2005. Measurements of bone mineral density of the radius in a medieval population. *Calcified Tissue International* 74 (2): 157–61.

Mehta, G., H. I. Roach, S. Langley-Evans, P. Taylor et al. 2002. Intrauterine exposure to a maternal low protein diet reduces adult bone mass and alters growth plate morphology in rats. *Calcified Tissue International* 71: 493–98.

Murphy, S., K. T. Khaw, H. May, and J. E. Compston. 1994. Parity and bone mineral density in middle-aged women. *Osteoporosis International* 4: 162–66.

Namgung, R., and R. C. Tsang. 2000. Factors affecting newborn bone mineral content: In utero effects on newborn bone mineralization. *Proceedings of the Nutrition Society* 59: 55–63.

Naylor, K. E., P. Iqbal, C. Fledelius, R. B. Fraser et al. 2000. The effect of pregnancy on bone density and bone turnover. *Journal of Bone and Mineral Research* 15: 129–37.

Nelson, D. A. 1984. Bone density in three archaeological populations. *American Journal of Physical Anthropology* 63: 198.

Nelson, D. A., and M. L. Villa. 2003. Ethnic differences in bone mass and architecture. In *Bone Loss and Osteoporosis: An Anthropological Perspective,* ed. S. C. Agarwal and S. D. Stout, 47–58. New York: Kluwer Plenum Academic Press.

Nelson, D. A., N. Sauer, and S. C. Agarwal. 2002. Evolutionary aspects of bone health. *Clinical Reviews in Bone and Mineral Metabolism* 1 (3): 169–79.

NOF (National Osteoporosis Foundation). 2006. http://www.nof.org/.

Nordin, B. E. C., B. G. Need, H. A. Morris, M. Horowitz et al. 1995. Bad habits and bad bones. In *Nutritional Aspects of Osteoporosis '94,* Challenges of Modern Medicine vol. 7, ed. P. Burckhardt and R. P. Heaney, 1–25. Rome: Ares-Serono Symposia.

Oreffo, R. O. C., B. Lashbrooke, H. I. Roach, N. M. P. Clarke, and C. Cooper. 2003. Maternal protein deficiency affects mesenchymal stem cell activity in the developing offspring. *Bone* 33: 100–107.

Ortner, D. J., W. Butler, J. Cafarella, and L. Milligan. 2001. Evidence of probable scurvy in subadults from archaeological sites in North America. *American Journal of Physical Anthropology* 114: 343–51.

Ortner, D. J., E. H. Kimmerle, and M. Diez. 1999. Probable evidence of scurvy in subadults from archeological sites in Peru. *American Journal of Physical Anthropology* 108: 321–31.

Orwoll, E. S. 1992. The effects of dietary protein insufficiency and excess on skeletal health. *Bone* 13: 343–50.

Palacios, C. 2006. The role of nutrients in bone health, from A to Z. *Critical Reviews in Food Science and Nutrition* 46 (8): 621–28.

Parfitt, A. M. 1990. Osteomalacia and related disorders. In *Metabolic Bone Diseases and Clinically Related Disorders*, 2nd ed., ed. L. V. Arioli and S. M. Krane, 329–96. Philadelphia: W. B. Saunders.

Parfitt, M. 2003. New concepts of bone remodeling: A unified spatial and temporal model with physiologic and pathophysiologic implications. In *Bone Loss and Osteoporosis: An Anthropological Perspective*, ed. S. C. Agarwal and S. D. Stout, 3–15. New York: Kluwer Plenum Academic Press.

Parsons, T. J., M. Van Dusseldorp, M. Van Der Vliet, K. Van De Werken et al. 1997. Reduced bone mass in Dutch adolescents fed a macrobiotic diet in early life. *Journal of Bone and Mineral Research* 12: 1486–94.

Pearce, M. S., F. N. Birrell, R. M. Francis, D. J. Rawlings et al. 2005. Lifecourse study of bone health at age 49–51 years: The Newcastle thousand families cohort study. *Journal of Epidemiology and Community Health* 59: 475–80.

Pearson, O. M., and D. E. Lieberman. 2004. The aging of Wolff's "Law": Ontogeny and responses to mechanical loading in cortical bone. *American Journal of Physical Anthropology* S47: 63–99.

Pearson, D., M. Kaur, P. San, N. Lawson et al. 2004. Recovery of pregnancy mediated bone loss during lactation. *Bone* 34 (3): 570–78.

Pfeiffer, S., and P. King. 1981. Intracortical bone remodeling and decrease in cortical mass among prehistoric Amerindians. *American Journal of Physical Anthropology* 54: 262.

Pfeiffer, S. K., and R. A. Lazenby. 1994. Low bone mass in past and present aboriginal populations. In *Advances in Nutritional Research*, vol. 9., ed. H. H. Draper, 35–51. New York: Plenum Press.

Poulsen, L.W., D. Qvesel, K. Brixen, A. Vesterby et al. 2001. Low bone mineral density in the femoral neck of medieval women: A result of multiparity? *Bone* 28 (4): 454–58.

Ralston, S. H. 2005. Genetic determinants of osteoporosis. *Current Opinion in Rheumatology* 17 (4): 475–79.

Rauch, F., and E. Schoenau. 2001. Changes in bone density during childhood and adolescence: An approach based on bone's biological organization. *Journal of Bone and Mineral Research* 16: 597–604.

Rewekant, A. 2001. Do environmental disturbances of an individual's growth and development influence the later bone involution processes? A study of two mediaeval populations. *International Journal of Osteoarchaeology* 11 (6): 433–43.

Richman, E. A., D. J. Ortner, and F. P. Schulter-Ellis. 1979. Differences in intracortical bone remodeling in three Aboriginal American populations: Possible dietary factors. *Calcified Tissue International* 28: 209–14.

Rosen, C. J. 2002. Nutrition and bone health in the elderly. *Clinical Review in Bone and Mineral Metabolism* 1 (3–4): 249–60.

Schoenau, E., C. M. Neu, F. Rauch, and F. Manz. 2001. The development of bone strength at the proximal radius during childhood and adolescence. *Journal of Clinical Endocrinology & Metabolism* 86: 613–18.

Schuette, S. A., M. Hegsted, M. B. Zemel, and H. M. Linkswiler. 1981. Renal acid, urinary cyclic AMP, and hydroxyproline excretion as affected by level of protein, sulfur, amino acid, and phosphorus intake. *Journal of Nutrition* 111 (12): 2106–16.

Sievänen, H., L. Józsa, I. Pap, M. Järvinen et al. 2007. Fragile external phenotype of modern human proximal femur in comparison with medieval bone. *Journal of Bone and Mineral Research* 22: 537–43.

Snow, C. E. 1948. *Indian Knoll Skeletons of Site Oh2, Ohio County, Kentucky.* University of Kentucky Reports in Anthropology, IV (3): Part II. Lexington: University of Kentucky Press.

Sowers, M. 1996. Pregnancy and lactation as risk factors for subsequent bone loss and osteoporosis. *Journal of Bone and Mineral Research* 11: 1052–60.

Sowers, M., M. K. Clark, B. Hollis, R. B. Wallace et al. 1992. Radial bone mineral density in pre- and perimenopausal women: A prospective study of rates and risk factors for loss. *Journal of Bone and Mineral Research* 7: 647–57.

Sowers, M., G. Corton, B. Shapiro, M. L. Jannausch et al. 1993. Changes in bone density with lactation. *Journal of the American Medical Association* 269: 3130–35.

Sowers, M., M. Crutchfield, M. Jannausch, M., S. Updike, and G. Corton. 1991. A prospective evaluation of bone mineral change in pregnancy. *Obstetrics and Gynecology* 77: 841–45.

Sowers, M., D. Eyre, B. W. Hollis, J. F. Randolph et al. 1995. Biochemical markers of bone turnover in lactating and nonlactating postpartum women. *Journal of Clinical Endocrinology and Metabolism* 80: 2210–16.

Sowers, M. R., and D. A. Galuska. 1993. Epidemiology of bone mass in premenopausal women. *Epidemiologic Reviews* 15: 374–98.

Specker, B. L. 2004. Nutrition in pregnancy and lactation. In *Nutrition and Bone Health,* ed. M. F. Holick and B. Dawson-Hughes, 139–56. Totawa, NJ: Humana Press.

Stevenson, J. C., B. Lees, M. Devenport, M. P. Cust et al. 1989. Determinants of bone density in normal women: Risk factors for future osteoporosis? *British Medical Journal* 298: 924–28.

Stini, W. 2003. Bone loss, fracture histories, and body composition characteristics of older males. In *Bone Loss and Osteoporosis: An Anthropological Perspective,* ed. S. C. Agarwal and S. D. Stout, 105–16. New York: Kluwer Plenum Academic Press.

Stuart-Macadam, P. L. 1988. Rickets as an interpretive tool. *Journal of Paleopathology* 2: 33–42.

———. 1989. Nutritional deficiency diseases: A survey of scurvy, rickets, and iron-deficiency anemia. In *Reconstruction of Life from the Skeleton,* ed. M. Y. Iscan and A. R. Kennedy, 201–22. New York: Alan R. Liss, Inc.

Thompson, D. D., and M. Gunness-Hey. 1981. Bone mineral-osteon analysis of Yupik-Inupiak skeletons. *American Journal of Physical Anthropology* 55: 1–7.

Thompson, D. D., S. D. Posner, W. S. Laughlin, and N. C. Blumenthal. 1983. Comparison of bone apatite in osteoporotic and normal Eskimos. *Calcified Tissue International* 35: 392.

Thompson, D. D., E. M. Salter, and W. S. Laughlin. 1981. Bone core analysis of Baffin Island skeletons. *Arctic Anthropology* 18 (1): 87–96.

Turner, C. H. 2002. Biomechanics of bone: Determinants of skeletal fragility and bone quality. *Osteoporosis International* 13 (2): 97–104.

Turner-Walker, G., U. Syversen, and S. Mays. 2001. The archaeology of osteoporosis. *Journal of European Archaeology* 4: 263–68.

Van Gerven, D. P., J. R. Hummert, and D. B. Burr. 1985. Cortical bone maintenance and geometry of the tibia in prehistoric children from Nubia's Batn el Hajar. *American Journal of Physical Anthropology* 66: 272–80.

Velasco-Vazquez, J., E. Gonzalez-Reimers, M. Arnay-De-La-Rosa, N. Barros-Lopez et al. 1999. Bone histology of prehistoric inhabitants of the Canary Islands: Comparison between El Hierro and Gran Canaria. *American Journal of Physical Anthropology* 110 (2): 201–13.

Vieth, R. 2003. Effects of vitamin D on bone and natural selection of skin color: How much vitamin D nutrition are we talking about? In *Bone Loss and Osteoporosis: An Anthropological Perspective,* ed. S. C. Agarwal and S. D. Stout, 139–51. New York: Kluwer Plenum Academic Press.

Vogel, M., M. Hahn, P. Caselitz, J. Woggan et al. 1990. Comparison of trabecular bone structure in man today and an ancient population in Western Germany. In *Bone Morphometry,* ed. H. E. Takahashi, 220–23. Niigata, Japan: Nishimura.

Ward, J. A., S. R. Lord, P. Williams, K. Anstey et al. 1995. Physiologic, health and lifestyle factors associated with femoral neck bone density in older women. *Bone* 16: 373S–378S.

Watts, N. B. 2002. Bone quality: Getting closer to a definition. *Journal of Bone and Mineral Research* 17 (7): 1148–50.

Weaver, D. S. 1998. Osteoporosis in the bioarchaeology of women. In *Sex and Gender in Paleopathological Perspective,* ed. A. Grauer and P. Stuart-Macadam, 27–46. Cambridge: Cambridge University Press.

Yano, K., L. K. Heilbrun, R. D. Wasnich, J. H. Hankin et al. 1985. The relationship between diet and bone mineral content of multiple skeletal sites in elderly Japanese-American men and women living in Hawaii. *American Journal of Clinical Nutrition* 42: 877–88.

• 10 •
Obesity
An Emerging Epidemic—
Temporal Trends in North America

P. T. Katzmarzyk

Introduction

Contemporary Western society places a premium on efficiency, and efforts to maximize efficiency have proliferated to virtually all aspects of our lives. For example, a large fraction of the technological devices that are marketed to North Americans are related in one way or another to making life "easier." The invention and uptake of remote controls, escalators, moving sidewalks, cell phones, computers, and other time-saving devices has largely engineered physical activity out of our lives. This drive towards a sedentary existence, coupled with an environment where affordable, energy-dense diets prevail, has led to a society in which it is more normal to be overweight than it is to be "normal" weight.

This chapter explores the trends in overweight and obesity that have occurred in North America in recent decades, and provides some insights into the burden that these changes have had on the health of the population. The current definitions and classification of body weight status are discussed first to provide background and context for the rest of the chapter.

Classification of Overweight and Obesity

The terms *overweight* and *obesity* refer to conditions of excess adiposity. The quantity and distribution of adipose tissue can be measured quite precisely in the laboratory using imaging techniques such as computed tomography, magnetic resonance imaging, or dual-energy X-ray absorptiometry (Heymsfield et al. 2005). These methods analyze the attenuation of x-rays or absorption of energy in specific body depots. Although highly accurate, these quantification methods are expensive and involve the use of bulky laboratory equipment. Imaging methods are often used to validate anthropometric field measures of obesity.

The precise measurement of adipose tissue is not always feasible in field studies or in large epidemiological investigations. Epidemiologists and human biologists generally rely on simple anthropometric markers of adiposity for the population surveillance of overweight and obesity. The most commonly used indicator of adiposity in both children and adults is the body mass index [BMI = weight (kg) / height (m^2)]. The BMI is moderately correlated with more direct measures of body fatness (Roche et al. 1981), and is easily calculated from height and weight, two measurements that are routinely made in the clinical or field setting.

Table 10.1 presents the current classification of body weight status in adults using the BMI. These criteria have been adopted by the World Health Organization (World Health Organization 1998), US National Institutes of Health (NIH 1998), and Health Canada (Health Canada 2003). Individuals can be classified as underweight (BMI < 18.5 kg/m^2), normal weight (BMI 18.5–24.9 kg/m^2), overweight (BMI 25–29.9 kg/m^2), or obese (BMI ≥ 30 kg/m^2). In addition, there are three classes of obesity (I, II, III) that correspond to increasing obesity-related health risks. These adult BMI categories should not be used for screening children and adolescents, as the BMI changes throughout normal growth and maturation.

There are currently two main approaches to the classification of body weight status using the BMI in children and adolescents: a distributional approach, and a

Table 10.1. Classification of body weight status in adults using the body mass index (BMI).

Weight Classification	BMI (kg/m^2)	Risk of Health Problems
Underweight	< 18.5	Increased
Normal weight	18.5–24.9	Least
Overweight	25.0–29.9	Increased
Obese	≥ 30	
Class I	30.0–34.9	High
Class II	35.0–39.9	Very high
Class III	≥ 40	Extremely high

Adapted from US National Institutes of Health and Health Canada Guidelines.

quasi-criterion referenced approach. Using a distributional approach, the US Centers for Disease Control and Prevention (CDC) have developed BMI percentiles for the classification of "risk of overweight" (BMI ≥ 85th and < 95th percentile) and "overweight" (BMI ≥ 95th percentile) (Kuczmarski et al. 2002). On the other hand, Cole and colleagues (Cole et al. 2000) have developed BMI thresholds for "overweight" and "obesity" that correspond to the adult thresholds of 25 kg/m^2 and 30 kg/m^2, respectively (Table 10.2). These thresholds were developed under the auspices of the International Obesity Task Force (IOTF) and are recommended for international comparisons of overweight and obesity prevalences (International Obesity Task Force 2004). Estimates of the prevalence of overweight and obesity in children and adolescents differ depending on whether the CDC or IOTF BMI thresholds are used; thus, it is important to specify which classification system was used (Flegal et al. 2001). The CDC BMI thresholds are more commonly used in clinical settings, as it is based on the same reference data as is used for stature and body mass. On the other hand, the IOTF thresholds are more often used in the surveillance of overweight and obesity levels, as it was developed in an international sample and is recommended by the IOTF for international comparisons. It is recommended that for comparison purposes, that overweight and obesity prevalences should be presented using both thresholds in scientific communications.

In addition to BMI, abdominal or waist circumference is also a useful marker of obesity-related health risk. Waist circumference has been shown to be the single best anthropometric predictor of abdominal visceral adipose tissue in adults (Pouliot et al. 1994; Rankinen et al. 1999). Waist circumference is also highly correlated with CT-measured intra-abdominal adipose tissue ($r = 0.84$) and subcutaneous abdominal adipose tissue ($r = 0.93$) (Goran et al. 1998) in children aged four to ten years, and it is effective as a screening tool for identifying children and youth aged three to nineteen with high DEXA-derived estimates of trunk fat mass (Taylor et al. 2000). There are currently several standard protocols for the measurement of waist circumference, but no consensus on the best technique. Common methods include measurement at the 1) level of iliac crest, 2) level of the lowest ribs, 3) midway between the lowest ribs and the iliac crest, 4) minimal waist, and 5) umbilicus. Waist circumference is a highly reproducible measurement; however, values measured at different sites can differ by up to 4 to 5 cm (Wang et al. 2003). Among adults, waist circumference values above 80 cm and 94 cm in women and men, respectively, indicate increased risk, and values above 88 cm and 102 cm respectively, indicate substantially increased risk (NIH 1998). Among children and youth, reference data have been developed for both Canada (Katzmarzyk 2004) and the United States (Fernandez et al. 2004); however, current recommendations do not advocate for the measurement of waist circumference as part of periodic screening for obesity in children and youth (American Academy of Pediatrics Committee on Nutrition 2003; International Obesity Task Force 2004; Katzmarzyk et al. 2007).

Table 10.2. International classification of body weight status among children and youth using the body mass index.

	Body mass Index			
	25 kg/m^2		30 kg/m^2	
Age (y)	Males	Females	Males	Females
18	25	25	30	30
17.5	24.73	24.85	29.70	29.84
17	24.46	24.70	29.41	29.69
16.5	24.19	24.54	29.14	29.56
16	23.90	24.37	28.88	29.43
15.5	23.60	24.17	28.60	29.29
15	23.29	23.94	28.30	29.11
14.5	22.96	23.66	27.98	28.87
14	22.62	23.34	27.63	28.57
13.5	22.27	22.98	27.25	28.20
13	21.91	22.58	26.84	27.76
12.5	21.56	22.14	26.43	27.24
12	21.22	21.68	26.02	26.67
11.5	20.89	21.20	25.58	26.05
11	20.55	20.74	25.10	25.42
10.5	20.20	20.29	24.57	24.77
10	19.84	19.86	24.00	24.11
9.5	19.46	19.45	23.39	23.46
9	19.10	19.07	22.77	22.81
8.5	18.76	18.69	22.17	22.18
8	18.44	18.35	21.60	21.57
7.5	18.16	18.03	21.09	21.01
7	17.92	17.75	20.63	20.51
6.5	17.71	17.53	20.23	20.08
6	17.55	17.34	19.78	19.65
5.5	17.45	17.20	19.47	19.34
5	17.42	17.15	19.30	19.17
4.5	17.47	17.19	19.26	19.12
4	17.55	17.28	19.29	19.15
3.5	17.69	17.40	19.39	19.23
3	17.89	17.56	19.57	19.36
2.5	18.13	17.76	19.80	19.55
2	18.41	18.02	20.09	19.81

Adapted from Cole et al. 2000.

Evolutionary Perspectives

There is consistent evidence that the prevalence of obesity has increased in many developed nations in recent decades. However, there is little information about the changes in BMI that have occurred over longer time periods. It is generally assumed that the high levels of adiposity observed today are unprecedented in history.

There is limited evidence from the fossil record to provide estimates of BMI along the early hominid lineage. In order to get a rough estimate, I have estimated the average BMI of early hominids using published estimates of stature (McHenry 1991) and body mass (McHenry 1992; Ruff and Walker 1993) for males. These early hominid estimates are compared to more recent samples, including estimated average BMI of British male army recruits (aged twenty-one to twenty-four years) from 1860–1863 (Rosenbaum 1988), and directly measured BMI values for representative samples of adult males in the United States from 1960–2000 (Okosun et al. 2004). The results of this analysis are presented in Figure 10.1, along the human evolutionary time scale.

Much of the evolution of *Australopithecus* and *Homo* over the long term has been characterized by a BMI on the order of 19 to 21 kg/m^2, which corresponds to the "normal weight" range for modern humans. Not surprisingly, there have been dramatic increases in BMI observed over the last couple of decades (an extremely small

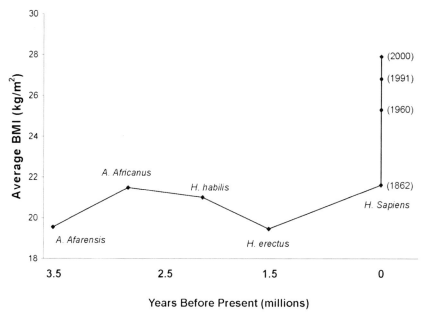

Figure 10.1. Estimated average body mass index in males across the evolution of hominids from *Australopithecus afarensis* to contemporary *Homo sapiens*.

fraction of the human evolutionary time scale). Indeed, for most of our evolution we have been hunter-gatherers (3–4 million years), while recent advances in agriculture and technology have occurred over a comparatively short time frame (~10,000 years). Thus, some scholars have argued that our evolutionary past set biological conditions for the recent meteoric rise in obesity (Brown and Konner 1987; Lev-Ran 2001; Ulijaszek and Lofink 2006; see also Leonard, Robertson, and Snodgrass, this volume).

The crux of this argument is that hominids in our evolutionary history have likely undergone natural selection for genetic traits that promote energy storage and reduce expenditure (Ulijaszek and Lofink 2006). Although, many mammals have the ability to overeat and increase adiposity (Uliaszek and Lofink 2006), *Homo*, in particular, may excel at it because of the need to store energy for a much larger brain size (see Leonard, Robertson, and Snodgrass, this volume). As well, fat storage is selectively advantageous to humans in terms of the reproduction and survival of offspring. For example, human females, relative to their nonhuman primate counterparts, lay down healthy amounts of fat before and during pregnancy in order to subsidize lactation, which is high in terms of daily energy expenditure (see Sellen, this volume). Finally, a number of scholars have suggested that seasonal food shortages throughout human history have led to natural selection that favored individuals who could effectively store energy during food deprivation (Neel 1962; Neel, Weder, and Julius 1998; Brown and Konner 1987). Alternatively, as argued by Leonard, Robertson, and Snodgrass (this volume), the rise in obesity in recent years may predominantly be a case of an imbalance in energy expenditure relative to energy intake due to our species' increasingly sedentary "modern" lifestyles.

In 2000 the average BMI for men peaked at 27.9 kg/m^2 in the United States, and past trends suggest that mean BMI levels will continue to increase (Okosun et al. 2004). It is clear that the recent cultural and technological changes in North American society have caused profound shifts in the biology of *Homo sapiens*. These shifts in the average BMI of the population are undoubtedly having an impact on the health of the population, and it has recently been estimated that obesity is negatively influencing life expectancy in the United States (Olshansky et al. 2005).

Obesity in Contemporary North America

There is consistent evidence that the prevalence of obesity has increased in North America in recent decades. Figure 10.2 presents the provincial/state-level prevalences of adult obesity in 1994 and 2005 in Canada and the United States. These data are from the Behavioral Risk Factor Surveillance System in the United States (Centers for Disease Control and Prevention 2006; Mokdad et al. 1999) and from the National Population Health Surveys/Canadian Community Health Surveys in Canada (Katzmarzyk 2002; Statistics Canada 2006). The values reported in these surveys are based on self-reported height and weight to estimate BMI and the prevalence of obesity. Given that respondents tend to underestimate their weight and overestimate

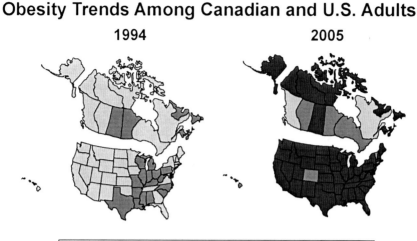

Figure 10.2. Prevalence of self-reported obesity (BMI ≥ 30 kg/m^2) in the United States and Canada, 1994 and 2005. Data from the United States from (Centers, 2006; Mokdad et al., 1999); data for Canada from (Katzmarzyk, 2002; Statistics Canada, 2006).

their height, which results in underestimates of their actual BMI (Connor Gorber et al. 2007), the absolute prevalences of obesity in Figure 10.2 should be considered conservative. Despite regional differences in absolute prevalence, there have been dramatic increases in the prevalence of adult obesity in every region of Canada and the United States in recent years.

Temporal Trends in Canada

In Canada, the current surveillance of obesity is conducted by Statistics Canada using the Canadian Community Health Surveys (CCHS). The CCHS are a series of cross-sectional surveys that are designed to provide cross-sectional estimates of health determinants, health status, and health system utilization across Canada (Statistics Canada 2003). The main component of the CCHS uses self-reported height and weight to assess the prevalence of obesity; however, both the 2004 CCHS Nutrition subcomponent and the 2005 CCHS included subsamples of the population with measured height and weight of sufficient size to provide representative national estimates of the prevalence of obesity. In 2005, the prevalence of obesity among adults based on self-reported BMI was 15.5 percent, while the measured prevalence was 24.3 percent (Statistics Canada 2006). The trends in adult obesity using both the self-reported and measured obesity data in Canada are shown in Table 10.3. It should be noted that the temporal trends are similar whether one relies on self-reported or

Table 10.3. Temporal trends in the prevalence of measured obesity (BMI ≥ 30 kg/m²) in Canada and the United States (1972-2005).

Year	Prevalence of Obesity (%)	
	Measured	Self-Reported
1972	11	n/a
1979	13.8	n/a
1985	n/a	6.2
1989	14.8	n/a
1990	n/a	9.7
1994	n/a	13.1
1996	n/a	12.5
1998	n/a	14.4
2000	n/a	14.8
2003	n/a	15.2
2004	23.1	n/a

measured obesity data; however, the absolute prevalence of obesity is higher at each time point when measured data are available. These results highlight the importance of using directly measured heights and weights in the estimation of the population prevalence of obesity.

The temporal trends in childhood and youth obesity in Canada have mirrored those seen among adults (Tremblay, Katzmarzyk, and Willms 2002). The most recent data using measured BMI and the IOTF thresholds (Cole et al. 2000) among children and youth aged two to seventeen years indicate that the prevalence of obesity has increased from approximately 3 percent in 1978-79 to 8 percent in 2004 (Shields 2006). Likewise, the percentage of children and youth classified as overweight (including obesity) increased from 15 percent to 26 percent over the same time period (Shields 2006). The trends over time in overweight and obesity were similar among boys and girls.

There are very little nationally representative data on waist circumference in Canada. Waist circumference was measured as part of the 1981 Canada Fitness Survey (Shephard 1986), the 1988 Campbell's Survey on Well-being (Stephens and Craig 1990), and the 1986-1992 Canadian Heart Health Surveys (MacLean et al. 1992). The changes in the prevalence of high waist circumference in men and women in Canada across these surveys are presented in Table 10.4. Similar to the trends observed for BMI, there have been increases over time in the proportion of Canadian adults who are at increased health risk due to abdominal obesity. The extent to which

Table 10.4. Prevalence of abdominal obesity (waist circumference > 88 cm in women; > 102 cm in men) among participants in the 1981 Canada Fitness Survey, 1988 Campbell's Survey of Well-being, and the 1986–1992 Canadian Heart Health Surveys.

	Prevalence of Abdominal Obesity (%)	
Year	Males	Females
1981	7.5	9.7
1988	11.6	14.6
1986–1992	14.1	19

changes in abdominal obesity have proceeded in concert with changes in overall weight for height have yet to be determined.

Temporal Trends in the United States

In the United States there are two main series of national surveys that collect data on obesity. The Behavioral Risk Factor Surveillance System (BRFSS) collects self-reported BMI data each year from a representative sample of the US population that is large enough to provide state-level prevalences (Centers for Disease Control and Prevention 2006). The US National Health and Nutrition Examination Survey (NHANES) collects data on measured height, weight, and waist circumference of the US population on an ongoing basis (Centers for Disease Control and Prevention 2007). It should be noted that data on obesity from the BRFSS and the NHANES are not directly comparable due to the differences in the data collection methods. For example, data from the 2000 BRFSS indicated that the self-reported prevalence of obesity was 19.8 percent (Mokdad et al. 2001), while directly measured data from the 1999–2000 NHANES indicated that the prevalence of obesity was 30.5 percent (Flegal et al. 2002).

The measured data from NHANES indicates that the prevalence of adult obesity has increased from approximately 10.5 percent in 1960 to 31.1 percent in 2003–04 in men and from approximately 15 percent to 33.2 percent in 2003–04 in women (Flegal et al. 1998; Ogden et al. 2006). Similar trends have been observed in American children and youth. In 2003–04, approximately 17 percent of children and youth aged two to nineteen had a BMI at or above the 95th percentile for BMI, while 33.6 percent of children and youth had a BMI at or above the 85th percentile for BMI (Ogden et al. 2006). In addition to the current high prevalence of overweight and obesity among US children and youth, there are strong indications that these prevalences have increased over time (Flegal and Troiano 2000).

Recent trends in adult waist circumference have been examined using data from NHANES III (1988–1994) and more recent cycles of the survey (Li et al. 2007). The age-adjusted trends in the means indicate that waist circumference has increased from 96 cm to 100.4 cm in men and from 89 cm to 94 cm in women between 1988–1994 and 2003–04. The corresponding changes in the prevalence of abdominal obesity (>88 cm in women; >102 cm in men) are 29.5 percent to 42.4 percent in men and 47 percent to 61.3 percent in women over the same time period (Li et al. 2007). These data suggest that the disease burden associated with abdominal obesity is also likely to increase as the increasing prevalence begins to affect the health of a larger proportion of the population.

Temporal Trends in Mexico

There are relatively little nationally representative data on the prevalence of overweight and obesity in Mexico. The data that are available consist of regional surveys and local studies, and several national surveys. The Mexican National Nutrition Surveys conducted in 1988 and 1999 provide a framework with which to examine temporal trends in obesity among women and children under age five (Rivera and Sepulveda Amor 2003). Data from the Mexican National Nutrition Surveys conducted between 1988 and 1999 also indicate a significant increase in the prevalence of obesity among women, increasing from 9 to 24 percent between the two surveys (Rivera et al. 2004). An increase in the prevalence of overweight (weight for length Z-score ≥2) was also observed among children under age five, increasing from 4.2 percent to 5.3 percent over the same time period (Rivera and Sepulveda Amor 2003).

Figure 10.3 illustrates the prevalences of measured obesity among adults in Mexico from the 1992–93 National Survey of Chronic Diseases (Arroyo et al. 2000) and the 2000 Mexican National Health Survey (Sanchez-Castillo et al. 2003). It is clear that there was an increase in the prevalence of obesity from 1992–93 to 2000 in both men and women. These surveys had different sampling frames and were not designed to provide continuing surveillance for obesity; however, the 2000 survey has a higher prevalence of obesity compared to the 1992–93 survey. Although reliable trend data using the current BMI thresholds for childhood obesity in Mexico are unavailable, data from the 2000 Mexican National Health Survey indicated that the prevalence of childhood overweight (including obesity) was quite high: 17 percent in boys and 20.6 percent on girls aged ten to seventeen years (Rio-Navarro et al. 2004).

The Public Health Burden of Obesity

Although many people view obesity mainly as an aesthetic issue, there are numerous health concerns associated with excess adiposity, including heart disease, stroke, type-2 diabetes, osteoarthritis, gall bladder disease, and several cancers (NIH 1998). Given the recent increases in the prevalence of obesity, these health risks translate into a substantial public health burden. Briefly, the public health burden of obesity refers to its impact on the public health infrastructure of a region or a country. There are several ways to estimate the public health burden associated with a particular risk factor or condition, including calculating its economic costs, morbidity burden,

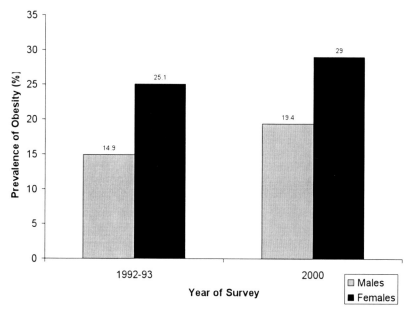

Figure 10.3. Temporal trends in the prevalence of measured obesity (BMI ≥ 30 kg/m²) among adults in Mexico from the 1992–93 National Survey of Chronic Diseases (Arroyo et al., 2000) and the 2000 Mexican National Health Survey (Sanchez-Castillo et al., 2003).

or the number of attributable deaths. The main statistic used to estimate the public health burden of a risk factor or condition is the Population Attributable Risk (PAR) percent, which is a theoretical estimate of the impact of a risk factor on society. Various equations have been developed for the PAR percent; however, the most popular is as follows:

$$\text{PAR percent} = [P(RR-1)] / [1 + P(RR-1)]$$

where P is the population prevalence of a risk factor, and RR is the relative risk of developing a particular disease for an individual given the presence of the risk factor. Thus, the PAR percent is a reflection of how common a risk factor is, and how "bad" it is at the individual level. Risk factors that have small effects on disease may have a major impact on public health if they are very common. Estimates of the public health burden associated with obesity are useful in highlighting its importance from a health perspective.

A recent meta-analysis of prospective longitudinal studies has quantified the relative risks and PAR percent associated with the major obesity-related diseases in Canada (Katzmarzyk and Janssen 2004). Table 10.5 presents the major chronic diseases associated with obesity, along with the summary relative risk estimates from the meta-analysis. The results indicate that the disease burden associated with obesity is

Table 10.5. Summary relative risk estimates and population attributable risks (PAR%) for obesity in Canada.

Disease	Summary RR	95% CI	PAR%
Coronary Artery Disease	2.24	2.04–2.45	15.4
Stroke	1.50	1.28–1.77	6.8
Hypertension	4.50	4.15–4.84	34.0
Colon Cancer	1.45	1.23–1.71	6.2
Postmenopausal Breast Cancer	1.47	1.40–1.54	6.5
Type-2 Diabetes	3.73	3.45–4.06	28.6
Gall Bladder Disease	3.33	2.86–3.85	25.5
Osteoarthritis	1.99	1.76–2.24	12.7

Adapted from Katzmarzyk and Janssen 2004.

significant, ranging from approximately 6 percent of colon cancer to 34 percent of hypertension. Stated another way, 6 percent of colon cancer cases and 34 percent of hypertension cases would theoretically disappear if obesity could be eliminated in Canada. Given recent trends, it is clear that obesity will not be completely eradicated in coming years; however, reducing the prevalence will see concomitant decreases in the public health burden.

There have been limited attempts to quantify the economic burden associated with obesity in North America. In order to estimate the economic burden of obesity, the PAR percent values are applied to the known direct and indirect costs for a particular disease. Thus, it is assumed that if obesity accounts for 34 percent of hypertension cases, it would theoretically account for 34 percent of the costs of treating hypertension. In Canada, Birmingham and colleagues estimated the direct medical costs of obesity in 1997 at $1.8 billion (Birmingham et al. 1999). More recently, it has been estimated that the total economic cost of obesity was $4.3 billion in 2001 (Katzmarzyk and Janssen 2004). Of the total cost, $1.6 billion was accounted for by direct medical expenses attributable to obesity, which include physician costs, hospital costs, drugs, and other medical costs. On the other hand, $2.7 billion was attributable to indirect costs, which include the value of years lost due to premature mortality and the value of activity days lost due to short-term and long-term disability. Table 10.6 outlines where the associated costs of obesity are distributed. It is clear that obesity accounts for a significant portion of health care dollars annually. Efforts to reduce the prevalence of obesity will reduce the associated economic burden.

Estimates from the United States indicate that the economic cost associated with obesity was approximately $70 billion in 1995, or 7 percent of the total health care costs in the country (Colditz 1999). A recent review of economic studies conducted

Table 10.6. Direct and indirect costs (CAD millions) of major chronic diseases associated with obesity in Canada, 2001.

Disease	Direct Costs	Indirect Costs	Direct cost attributable to obesity	Indirect cost attributable to obesity
Coronary Artery Disease	2429.6	6296.0	374.6	970.7
Stroke	1691.5	1458.4	115.8	99.9
Hypertension	1530.2	1352.9	519.8	459.6
Colon Cancer	278.9	1331.9	17.3	82.6
Post-menopausal Breast Cancer	350.1	1671.5	22.6	130.6
Type 2 Diabetes	800.8	588.7	229.3	168.6
Gall Bladder Disease	691.4	452.0	176.4	115.3
Osteoarthritis	1121.3	5814.4	142.5	738.7
Total			1598.3	2743.4

Adapted from Katzmarzyk and Janssen 2004.

around the world indicated that between 2 percent and 7 percent of health care costs in individual countries are directly attributable to obesity (Thompson and Wolf 2001).

In addition to the economic costs associated with obesity, several investigators have estimated the attributable mortality associated with obesity. The relative risks of premature mortality increase across normal weight, overweight, and increasing classes of obesity. Results from a thirteen-year follow-up from the 1981 Canada Fitness Survey demonstrated relative risks of 1.16 in overweight, 1.25 in obese class I, and 2.96 in obese class II and III men and women for all-cause mortality by comparison to the normal weight reference group (Katzmarzyk and Ardern 2004). The authors estimated that 9.4 percent of all premature deaths (among adults aged twenty to sixty-four years) are directly attributable to overweight and obesity in Canada. In 2000, this amounted to approximately 4300 deaths due to overweight and obesity. By comparison, recent estimates for the United States range from 111,909 to 365,000 obesity-related deaths per year (Allison et al. 1999; Flegal et al. 2005; Mokdad et al. 2004). Differences in study design and the age ranges studied account for the differences across studies. For example, studies that include the entire adult age range in the analysis yield higher estimates of mortality given that most of the deaths in North America occur after age seventy. Although there are differences in the methodologies employed among studies and there is debate about the most appropriate analytical approach, it is clear that obesity is taking a mounting toll on the lives of North Americans.

Conclusion

In summary, there is consistent evidence that human obesity in North America is at an unprecedented level. This high prevalence is beginning to put undue stress on public health—one that is likely to continue to increase unless the growing prevalence can be halted. Given that the increases in childhood obesity have mirrored those seen among adults, this is cause for concern. There is a tendency for overweight and obese children to become obese adults (Whitaker et al. 1997); thus, the problem of adulthood obesity is likely to be compounded in the coming years as a direct consequence of the increases in childhood obesity. A combined approach of physical activity, dietary, and ecological interventions is warranted in the primary prevention of obesity. Aggressive public health campaigns are required to slow or reverse the recent temporal trends in obesity.

References

Allison, D. B., K. R. Fontaine, J. E. Manson, J. Stevens et al. 1999. Annual deaths attributable to obesity in the United States. *Journal of the American Medical Association* 282: 1530–38.

American Academy of Pediatrics Committee on Nutrition. 2003. Prevention of pediatric overweight and obesity. *Pediatrics* 112: 424–30.

Arroyo, P., A. Loria, V, Fernandez, K. M. Flegal et al. 2000. Prevalence of pre-obesity and obesity in urban adult Mexicans in comparison with other large surveys. *Obesity Research* 8: 179–85.

Birmingham, C. L., J. L. Muller, A. Palepu, J. J. Spinelli et al. 1999. The cost of obesity in Canada. *Canadian Medical Association Journal* 160: 483–88.

Brown, P. J., and M. Konner. 1987. An anthropological perspective on obesity. *Annals of New York Academy of Sciences* 499: 29–46.

Centers for Disease Control and Prevention. 2006. State-specific prevalence of obesity among adults—United States, 2005. *Morbidity and Mortality Weekly Reports* 55: 985–88.

Centers for Disease Control and Prevention. 2007. *National Health and Nutrition Examination Survey*. http://www.cdc.gov/nchs/nhanes.htm. Accessed 23 April 2007.

Colditz, G. A. 1999. Economic costs of obesity and inactivity. *Medicine and Science in Sports and Exercise* 31: S663–S667.

Cole, T. J., M. C. Bellizzi, K. M. Flegal, and W. H. Dietz. 2000. Establishing a standard definition for child overweight and obesity worldwide: International survey. *British Medical Journal* 320: 1240–43.

Connor Gorber, S., M. S. Tremblay, D. Moher, and B. Gorber. 2007. A comparison of direct vs. self-report measures for assessing height, weight and body mass index: A systematic review. *Obesity Reviews* 8: 307–26.

Fernandez, J. R., D. T. Redden, A. Pietrobelli, and D. B. Allison. 2004. Waist circumference percentiles in nationally representative samples of African-American, European-American, and Mexican-American children and adolescents. *Journal of Pediatrics* 145: 439–44.

Flegal, K. M., M. D. Carroll, R. J. Kuczmarski, and C. L. Johnson. 1998. Overweight and obesity in the United States: Prevalence and trends, 1960–1994. *International Journal of Obesity and Related Metabolic Disorders* 22: 39–47.

Flegal, K. M., M. D. Carroll, C. L. Ogden, and C. L. Johnson. 2002. Prevalence and trends in obesity among US adults, 1999–2000. *Journal of the American Medical Association* 288: 1723–27.

Flegal, K. M., B. I. Graubard, D. F. Williamson, and M. H. Gail. 2005. Excess deaths associated with underweight, overweight and obesity. *Journal of the American Medical Association* 293: 1861–67.

Flegal, K. M., C. L. Ogden, R. Wei, R. L. Kuczmarski et al. 2001. Prevalence of overweight in US children: Comparison of US growth charts from the Centers for Disease Control and Prevention with other reference values for body mass index. *American Journal of Clinical Nutrition* 73: 1086–93.

Flegal, K. M., and R. P. Troiano. 2000. Changes in the distribution of body mass index of adults and children in the US population. *International Journal of Obesity and Related Metabolic Disorders* 24: 807–18.

Goran, M. I., B. A. Gower, M. Treuth, and T. R. Nagy. 1998. Prediction of intra-abdominal and subcutaneous abdominal adipose tissue in healthy pre-pubertal children. *International Journal of Obesity and Related Metabolic Disorders* 22: 549–58.

Health Canada 2003. *Canadian Guidelines for Body Weight Classification in Adults.* Cat. No. H49-179/2003E. Ottawa, ON, Health Canada.

Heymsfield, S. B., T. G. Lohman, Z. Wang, and S. B. Going. 2005. *Human Body Composition.* 2nd ed. Champaign, IL: Human Kinetics.

International Obesity Task Force 2004. Assessment of obesity: Which child is fat? *Obesity Reviews* 5 (Suppl.): 10–15.

Katzmarzyk, P. T. 2002. The Canadian obesity epidemic, 1985–1998. *Canadian Medical Association Journal* 166: 1039–40.

———. 2004. Waist circumference percentiles for Canadian youth 11–18 y of age. *European Journal of Clinical Nutrition* 58: 1011–15.

Katzmarzyk, P. T., and C. I. Ardern. 2004. Overweight and obesity mortality trends in Canada, 1985–2000. *Canadian Journal of Public Health* 95: 16–20.

Katzmarzyk, P. T., and I. Janssen. 2004. The economic costs associated with physical inactivity and obesity in Canada: An update. *Canadian Journal of Applied Physiology* 29: 90–115.

Katzmarzyk, P. T., M. S. Tremblay, I. Janssen, and K. Morrison. 2007. Identification of overweight and obesity in children and adolescents: Canadian clinical practice guidelines on the management and prevention of obesity. *Canadian Medical Association Journal* 176 (8): 27–32.

Kuczmarski, R. J., C. L. Ogden, S. S. Guo, L. M. Grummer-Strawn et al. 2002. 2000 CDC growth charts for the United States: Methods and development. *Vital Health Statistics 11* 246: 1–190.

Lev-Ran, A. 2001. Human obesity: An evolutionary approach to understanding our bulging waistline. *Diabetes Metabolism Research Review* 17: 347–62.

Li, C., E. S. Ford, L. C. McGuire, and A. H. Mokdad. 2007. Increasing trends in waist circumference and abdominal obesity among US adults. *Obesity* 15: 216–24.

MacLean, D. R., A. Petrasovits, M. Nargundkar, P. W. Connelly et al. 1992. Canadian heart health surveys: A profile of cardiovascular risk. Survey methods and data analysis. *Canadian Medical Association Journal* 146: 1969–74.

McHenry, H. M. 1991. Femoral lengths and stature in Plio-Pleistocene hominids. *American Journal of Physical Anthropology* 85: 149–58.

———. 1992. Body size and proportions in early hominids. *American Journal of Physical Anthropology* 87: 407–31.

Mokdad, A. H., B. A. Bowman, E. S. Ford, F. Vinicor et al. 2001. The continuing epidemics of obesity and diabetes in the United States. *Journal of the American Medical Association* 286: 1195–200.

Mokdad, A. H., J. S. Marks, D. F. Stroup, and J. L. Gerberding. 2004. Actual causes of death in the United States, 2000. *Journal of the American Medical Association* 291: 1238–45.

Mokdad, A. H., M. K. Serdula, W. H. Dietz, B. A. Bowman et al. 1999. The spread of the obesity epidemic in the United States, 1991–1998. *Journal of the American Medical Association* 282: 1519–22.

Neel, J. 1962. Diabetes mellitus: A "thrifty" genotype rendered detrimental by "progress"? *American Journal of Human Genetics* 14: 353–62.

Neel, J. V., A. B. Weder, and S. Julius. 1998. Type II diabetes, essential hypertension, and obesity as "syndromes of impaired genetic homeostasis": The "thrifty genotype" hypothesis enters the 21st century. *Perspectives in Biology and Medicine* 42: 44–74.

NIH 1998. *Clinical Guidelines for the Identification, Evaluation, and Treatment of Overweight and Obesity in Adults.* Bethesda, MD: National Institutes of Health.

Ogden, C. L., M. D. Carroll, L. R. Curtin, M. A. McDowell et al. 2006. Prevalence of overweight and obesity in the United States, 1999–2004. *Journal of the American Medical Association* 295: 1549–55.

Okosun, I. S., K. M. D. Chandra, A. Boev, J. M. Boltri et al. 2004. Abdominal adiposity in U.S. adults: Prevalence and trends, 1960–2000. *Preventive Medicine* 39: 197–206.

Olshansky, S. J., D. J. Passaro, R. C. Hershow, J. Layden et al. 2005. A potential decline in life expectancy in the United States in the 21st century. *New England Journal of Medicine* 352: 1138–45.

Pouliot, M. C., J. P. Despres, S. Lemieux, S. Moorjani et al. 1994. Waist circumference and abdominal sagittal diameter: Best simple anthropometric indexes of abdominal visceral adipose tissue accumulation and related cardiovascular risk in men and women. *American Journal of Cardiology* 73: 460–68.

Rankinen, T., S. Y. Kim, L. Perusse, J. P. Despres et al. 1999. The prediction of abdominal visceral fat level from body composition and anthropometry: ROC analysis. *International Journal of Obesity and Related Metabolic Disorders* 23: 801–9.

del Rio-Navarro, B. E., O. Velazquez-Monroy, C. P. Sanchez-Castillo, A. Lara-Esqued et al. 2004. The high prevalence of overweight and obesity in Mexican children. *Obesity Research* 12: 215–23.

Rivera, J. A., S. Barquera, T. Gonzalez-Cossio, G. Olaiz et al. 2004. Nutrition transition in Mexico and in other Latin American countries. *Nutrition Reviews* 62: S149–S157.

Rivera, J. A., and J. Sepulveda Amor. 2003. Conclusions from the Mexican National Nutrition Survey 1999: Translating results into nutrition policy. *Salud Publica de Mexico* 45: S565–S575.

Roche, A. F., R. M. Siervogel, W. C. Chumlea, and P. Webb. 1981. Grading body fatness from limited anthropometric data. *American Journal of Clinical Nutrition* 34: 2831–38.

Rosenbaum, S. 1988. 100 years of heights and weights. *Journal of the Royal Statistical Society A* 151, Part 2: 276–309.

Ruff, C. B., and A. Walker. 1993. Body size and body shape. In *The Nariokotome Homo erectus Skeleton*, ed. A. Walker and R. Leakey, 234–65. Cambridge, MA: Harvard University Press.

Sanchez-Castillo, C. P., O. Velazquez-Monroy, A. Berber, A. Lara-Esqueda et al. 2003. Anthropometric cutoff points for predicting chronic diseases in the Mexican National Health Survey 2000. *Obesity Research* 11: 442–51.

Shephard, R. J. 1986. *Fitness of a Nation: Lessons from the Canada Fitness Survey.* New York: Karger.

Shields, M. 2006. Overweight and obesity among children and youth. *Health Reports* 17: 27–42.

Statistics Canada. 2003. CCHS Cycle 1.1 (2000–2001), Public use Microdata File Documentation. Ottawa, ON, Statistics Canada.

———. 2006. *Health Indicators*, vol. 1 Catalogue #82-221-XIE. Ottawa: Statistics Canada.

Stephens, T., and C. L. Craig 1990. *The Well-Being of Canadians: Highlights of the 1988 Campbell's Survey.* Ottawa: Canadian Fitness and Lifestyle Research Institute.

Taylor, R. W., I. E. Jones, S. M. Williams, and A. Goulding. 2000. Evaluation of waist circumference, waist-to-hip ratio, and the conicity index as screening tools for high trunk fat mass, as measured by dual-energy X-ray absorptiometry, in children 3–19 y. *American Journal of Clinical Nutrition* 72: 490–95.

Thompson, D., and A. M. Wolf. 2001. The medical-care cost burden of obesity. *Obesity Reviews* 2: 189–97.

Tremblay, M. S., P. T. Katzmarzyk, and J. D. Willms. 2002. Temporal trends in overweight and obesity in Canada, 1981–1996. *International Journal of Obesity and Related Metabolic Disorders* 26: 538–43.

Ulijaszek, S. J., and H. Lofink. 2006. Obesity in biocultural perspective. *Annual Review of Anthropology* 35: 337–60.

Wang, J., J. C. Thornton, S. Bari, B. Williamson et al. 2003. Comparisons of waist circumferences measured at 4 sites. *American Journal of Clinical Nutrition* 77: 379–84.

Whitaker, R. C., J. A. Wright, M. S. Pepe, K. D. Seidel et al. 1997. Predicting obesity in young adulthood from childhood and parental obesity. *New England Journal of Medicine* 337: 869–73.

World Health Organization 1998. *Obesity: Preventing and Managing the Global Epidemic.* Report of a WHO Consultation on Obesity, Geneva, 3–5 June 1997. Geneva: World Health Organization.

CONCLUSION

● ● ●

Diet and Nutrition in Biocultural Perspective
Back to the Future

T. Prowse and T. Moffat

Introduction

We begin this chapter by addressing the challenge we set in the introduction of this book, to link studies of diet and nutrition in the past and present to inform and engage one another. Thus, the first part of this chapter is devoted to a critical discussion of some of the theoretical, topical, and methodological links found in this volume. The breadth displayed in the chapters demonstrates the diversity of approaches in the biocultural study of human diet and nutrition. We believe that there is room for even more diversity as well as a need to consider some common theoretical touchstones that may aid in connecting research. The second part of this chapter builds on ideas and themes developed in our contributors' chapters. We suggest some potential future research directions that are linked to but go beyond more traditional areas found in biocultural studies of diet and nutrition.

Past Meets Present

With the rise of postmodernism at the end of the 1980s, there was a growing awareness of how studies of the past are influenced by modern narratives about human nature (e.g., Landau 1991; Wiber 1998). Conversely, we utilize our understanding of our human origins and past life ways to contemplate our contemporary existence (Goodenough 2004; Ward 2003). This is exemplified by recent trends in dietary fads, such as the "Paleoprescription", that advocate the return to foods consumed by our hunter-gatherer ancestors, before the advent of agriculture and a reliance on a cereal-based diet, in order to prevent contemporary chronic diseases such as heart disease and diabetes (Eaton and Konner 1985). Though there are justifiable criticisms of this kind of scholarship (Nestle 2000), we argue that we should not reject analyses of the dialectic between diet and nutrition in the past and present. Indeed, whether we like it or not, they inform one another, and their intersections can be fruitful. As McElroy (1990) suggests, we need to recognize that there is a feedback loop between biology and culture, just as there is a reciprocal relationship between past and present. At the same time, we must critically examine the utility and validity of our models.

Sellen (chapter 3), for example, is explicit in his commitment to employing our evolutionary heritage as a context for examining contemporary infant and young child feeding practices. He contends that mammalian biology is essentially conservative, although there have been some unique human adaptations that have been beneficial to our species, inasmuch as our population growth has outstripped any other mammal on the planet. He argues that there is a universal "best practice" for infant and child feeding that we ignore at our peril. Sellen's viewpoint in this volume is comparable to the paleodiet argument, mentioned above, that our modern diet is "out of step" with that to which our bodies were adapted during tens of thousands of years of hunter-gatherer subsistence (O'Keefe and Cordain 2004). Other authors, however, simply point out how social and cultural factors can affect the biological aspects of human nutrition, and more specifically human lactation; for example, through the prescription of breastfeeding and weaning schedules in the ancient world (Dupras, chapter 4; Prowse et al., chapter 8), or shifting political-economic circumstances in twentieth-century Gibraltar (Sawchuk, Bryce, and Burke, chapter 5).

Leonard, Robertson, and Snodgrass (chapter 1) argue that the commitment to an omnivorous diet with a more significant amount of energy-dense meat in the diet of *Homo* was in part responsible for one of the most significant evolutionary changes in hominid evolution, that is, cerebral development and expansion. This is relevant to current nutritional debates about the health advantages and disadvantages of vegetarianism and omnivory. The "expensive tissue hypothesis" (Aiello and Wheeler 1995), for example, has emphasized animal foods exclusively as necessary for hominid brain expansion. Whereas others have argued that there are packages of energy-dense non-animal foods that may also have played this role, such as tubers (See Pennisi 1999 for a summary). Leonard (2002) stresses our omnivorous and flexible adaptations and notes that both meat and high-quality vegetable foods, as well as innovations such as cooking and food-sharing, may have been catalysts for the improved quality of diet at the time of brain expansion.

Katzmarzyk (chapter 10) reflects on the apparent lack of obesity in the past among our hominid ancestors and on whether contemporary challenges of obesity are in part rooted in our evolutionary heritage. Similarly, Leonard, Robertson, and Snodgrass in this volume link our evolutionary past to modern health issues, particularly obesity, and point out that it is not simply "input" (what we eat) but also "output" that counts; through our modernized lifestyles we have decreased our energy expenditure. Biocultural researchers have long realized that obesity is much more than eating too much and exercising too little (Ritenbaugh 1991). In order to fully understand modern nutritional health issues we need to frame them within our evolutionary past, and by using a biocultural lens we can examine the complex interconnections between biology and culture and remind ourselves that there are no simple explanations for major changes in our evolutionary history, or in contemporary health dilemmas.

What is significant about these chapters is that they make it clear that understanding our human origins and life histories in the deep past may well have a bearing on our present state and future evolution. That is not to say that we are biologically determined, but that as biological beings we inherit a suite of behaviors that we carry with us as a species.

Other chapters in this volume that investigate diet and nutrition in past populations rely on comparative knowledge from modern epidemiological and anthropological studies of child growth (Pfeiffer and Harrington, chapter 2; Prowse et al. chapter 8), infant and young child feeding (Dupras, chapter 4; Prowse et al., chapter 8), and osteoporosis (Agarwal and Glencross, chapter 9). Modern medical understanding of human nutrition and health informs the analysis of evidence observed in the skeletal record and, in turn, we suggest that studies of past populations resonate with and have implications for contemporary issues surrounding diet, nutrition and disease. Pfeiffer and Harrington's proposition that slow growth of Stone Age children in Southern Africa may have been an evolutionary adaptation to the environment is reminiscent of the "small but healthy hypothesis" first proposed by Seckler (1980) and subsequently critiqued by nutritionists and anthropologists (Beaton 1989; Martorell 1989; Messer, 1989). The "small but healthy hypothesis" proposed that children in contemporary, developing countries are beneficially adapted to lower quantities of food, particularly protein. Pfeiffer and Harrington's hypothesis, however, differs from the "small but healthy hypothesis" in that they suggest that the strong selective force for the Stone Age population with small body size was not a low-quality diet, but rather the ability to maneuver in a rocky and rugged physical environment.

Agarwal and Glencross's (chapter 9) review of studies of osteoporosis in past populations raises interesting questions about how we study this disease in contemporary settings. They argue that we should consider diet and nutrition in relation to osteoporosis in terms of bone degeneration through the life course, demonstrating once again the need to explore both biological and cultural variables to fully understand these issues. This is more difficult to do in contemporary studies requiring very detailed retrospective or long-term prospective surveys, and thus perspectives from studies of past populations are essential for the life course approach they advocate.

Studies of recent historical and contemporary populations in this volume address the imperative to consider local culture and ideology in tandem with political-economy to investigate human diet and nutrition. Sawchuk, Bryce, and Burke (chapter 5) demonstrate how women's participation in the labor force—as well as information dissemination on the part of the government, medical professionals, the media, and exposure to other infant feeding practices found in European nations—influenced the rise, fall, and then rise again of breastfeeding in twentieth-century Gibraltar. Casiday and colleagues' (chapter 7) examination of humanitarian food aid at the local level during a food crisis in Niger illustrates how people's existing strategies for sharing food at the household level is at odds with the humanitarian agencies' distribution strategies. While the food agencies seek to target the most vulnerable children—assessed quantitatively by anthropometrics—families seek to distribute the resources evenly among offspring, and household benefits may supersede the needs of individual children.

While a consideration of ideology exemplified in these contemporary case studies is much more challenging in studies of diet and nutrition in ancient and prehistoric populations, it is not impossible. Prowse and colleagues (chapter 8), for example, take the medical advice of Roman medical writers into account when considering infant and young child feeding practices among the people of Isola Sacra. Similarly, Dupras (chapter 4) uses ancient writers and written contracts to infer attitudes towards the timing of breastfeeding and weaning in Roman Egypt. These chapters also relate to Sellen's (chapter 3) in their examination of the social context within which decisions are made about infant and young child feeding practices.

Finally, Moffat and Finnis (chapter 6) in their investigation of dietary change in urban Nepal note that studies of dietary transitions among contemporary populations are rooted in a tradition of research on large-scale dietary transitions in archaeological and skeletal remains (e.g., Molleson, Jones, and Jones 1993; Leonard 1994; Cachel 1997; Schmidt 2001). What differs between the time frames is scale and methodology. Studies of contemporary dietary transitions are usually conducted after the transition has recently occurred, so compared to studies of the past it is difficult to get the same level of information about the health effects of the transition. Conversely, because the people in contemporary populations are still alive, one can obtain dietary data directly from them in order to do analyses of the quality of their diet.

Modern dietary transitions may have parallels to what occurred in the past. In Goodman and Armelagos's (2000) study of food distribution among Mississippian peoples at the Dickson Mounds in Illinois, they hypothesize that even though there were ample protein-rich foods available in a mixed economy of hunting and agriculture, people chose to trade high-quality foodstuffs for material goods, thereby lowering the nutritional status of the population. Thus, studies of the behavior of food distribution and consumption of contemporary people can be useful in providing ideas and models for analyzing strategies and practices among past peoples, and we may be able to infer quality of diet by looking at skeletal evidence of changes in morbidity and mortality. Like the shift from hunting and gathering to agriculture, the transition from subsistence-based agriculture to industrialization and the globalization of food is a profound change in the human subsistence economy that will continue to have huge ramifications for human diet and nutrition.

Future Directions

We now turn to a consideration of future directions in the biocultural study of diet and nutrition based on topics and themes introduced in the preceding chapters. We begin by thinking about evolutionary aspects of diet and nutrition and suggest some new areas of exploration. The first is an idea that picks up on Leonard, Roberson, and Snodgrass's (chapter 1) argument that dietary change was pivotal to human evolution. But rather than considering choice quality foods, such as meat, it is an exploration of what have been called "fallback foods." Fallback foods, or marginal foods, are of lower nutritional quality and relatively high abundance, but are important in times of food scarcity (Marshal and Wrangham 2007). Studies of living nonhuman primates have led researchers to hypothesize that fallback foods have a significant impact on primate physiology, foraging patterns, and social behavior (e.g., Conklin-Brittain, Wrangham, and Hunt 1998; White 1998; Yamakoshi 1998; Furuichi, Hashimoto, and Yashiro 2001; Lambert et al. 2004—see Marshall and Wrangham 2007 for a review). A consideration of the role of fallback foods in hominid evolution comes from comparisons of African apes with hominid fossils, as well as ethnographic analogy among contemporary human foragers (Ungar 2004; Laden and Wrangham 2005). Ungar (2004) argues that differences in dental topography between early australopithecines and later *Homo* species are of the same magnitude as those found between living gorillas and chimpanzees. Since these living primates consume different fallback foods in areas where the species overlap, Ungar (2004) suggests that early hominids relied on fallback foods that differed from those of the later *Homo* species. In applying this model to early hominids, Laden and Wrangham (2005) hypothesize that part of their adaptive shift to a nonhumid, arid grassland may have included a change in fallback foods, which they propose may have been underground storage organs (USOs) such as rhizomes, tubers, corms, bulbs, and caudices. These types of plants are adept at surviving in adverse growing conditions, and thus would have been most plentiful relative to other foods during long, dry seasons. There is much more work to be done on understanding the role of preferred foods and fallback foods in human evolution, particularly if tooth structure and function are affected by fallback foods, since this is often the only evidence we have of many early hominid species. So, it is not what they habitually ate, but rather what they had to eat in times of scarcity that may have had a greater impact on their physiology. This shift in thinking has important consequences for interpreting what early hominids were actually eating.

Again, building on the ideas of Leonard, Roberson, and Snodgrass (chapter 1), a further consideration of the role of cooking in human evolution is warranted. In a recent book called *Catching Fire: How Cooking Made us Human,* Richard Wrangham (2009) convincingly argues that cooking may have been more important in improving dietary quality than the increase in meat eating related to brain expansion and indeed a fundamental part of our evolutionary history as a species. As cooking is a uniquely human practice, and indeed appears to be a universal, it warrants further research both in nutritional science and anthropology.

Another area in the biocultural study of diet and nutrition that warrants more research involves the life history approach adopted by Sellen (chapter 3). Life history theory is a framework that explores biology and behavior in relation to stages of maturational development throughout the lifetime of an individual (e.g., weaning, reproductive maturity, senescence). This approach has been used in the study of life history evolution of living and fossils primate species (e.g., Schwartz et al. 2002, 2005; Dirks and Bowman 2007). In this volume, Sellen concentrates on how the inclusion of transitional feeding, unique to infant and young child feeding, may have been related to modifications in human life history by decreasing birth spacing and increasing fertility. This approach can be expanded to consider the role of other changes in hominid diet and nutrition that may have altered life history. Kaplan et al. (2000), for example, propose that the shift to acquiring energy-dense, large-package, skill-intensive food resources is responsible for the unique life history traits associated with *Homo,* including a long juvenile developmental period, an intergenerational system of resource flow, and a very long lifespan. More comparative primate studies as well as reanalysis of hunter-gatherer studies of diet and nutrition are required to test this hypothesis.

Linked to the life history approach is a greater understanding and appreciation of the early years of the life course. Seven of the chapters in this volume touch on or are mainly concerned with the diet and/or nutrition of infants and children. We argue that this is part of a larger trend in the anthropology of children and childhood (Panter-Brick 1998; Lewis 2007). To date much of the research on children's diet has been exclusively or mostly concerned with breastfeeding. We suggest that more research be pursued about children of weaning age and older to continue the investigation of diet and nutrition in later stages of the life course. The life course perspective recognizes that biological development is affected by social and historical context (Harlow and Laurence 2002), and this concept can be productively applied to both past and present populations. Life course analysis integrates nicely with the biocultural approach, because they both recognize the importance of historical time and place and social relations in the analysis of human diet and nutrition.

We know that children's diets are both quantitatively and qualitatively different from that of adults due to their unique biology and social status, yet they are linked to the rest of the life course, because early developmental stress can impact later morbidity and mortality (Kuzawa 2005). There is a dearth of studies of children's diet and nutrition in both the past and present, so we suggest that future studies explicitly acknowledge they are studying children's diets as unique from their adult counterparts, and recognize this as a distinct stage in the life course. These might include perspectives from adults about the household distribution of food (see, for example, Casiday et al. chapter 7), as well as an examination of children's agency in food choices and food foraging and acquisition (Hawkes, O'Connell, and Blurton 1995; Panter-Brick, Todd, and Baker 1996; Bird and Bliege Bird 2000). In the postindustrial context, children's food choices include what is known by marketers as "pester power": children are often strongly influenced by food advertising in their attempts to persuade their caregivers to purchase preferred foods (Arnas 2006). This behavior has implications for modern trends in child nutrition and obesity.

An important methodological tool to investigate children's diets in archaeological samples is isotopic analysis, which is now customary. As we have just mentioned, children are not "little adults," and too few studies investigate the interval between the postweaning period and "adulthood." Dupras (chapter 4) and Prowse et al. (chapter 8) demonstrate, however, that we can access dietary information from those individuals who survived weaning and early childhood. Dietary studies of past human groups may benefit from new developments in isotopic research. Birchall et al. (2005) report that hydrogen isotopes display a strong correlation with trophic level (similar to $\delta^{15}N$), particularly among terrestrial consumers. The additional evidence from hydrogen isotopes may help to further elucidate dietary patterns in the past, potentially including patterns of breastfeeding and weaning. Further developments in the isotopic investigation of past diet include the analysis of specific amino acids from bone collagen, which may be able to discern more detailed dietary information than bulk protein alone (Fogel and Tuross 2003; Corr et al. 2005; McCullagh, Juchelka, and Hedges 2006). These areas of research have the potential to greatly expand our knowledge about diet in archaeological contexts.

The rise in the global prevalence of obesity is profoundly changing the physical shape and morbidity and mortality profile of the human species (Katzmarzyk, chapter 10). This is in part due to the industrial and postindustrial mode in which we currently produce, distribute, and process our food. We have created a food system to feed the billions of humans living on our planet, resulting in foods laced with pesticides, growth-promoting hormones, antibiotics, high fructose corn syrup, sodium, and trans fats, to name a few offending ingredients. As Michael Pollan, in his popular book *The Omnivore's Dilemma* (2006), notes, as flexible omnivores who are able to consume and thrive on a wide variety of foods, in modern times, as in our evolutionary past, we are faced with the challenge of trying to discern those foods that are palatable and nutritious from those that are toxic. As anthropologists we need to consider this new industrial/postindustrial food system as the third great shift in subsistence, after hunting and gathering and the rise of agriculture. Just as we have done for the previous two modes of subsistence, we must consider the profound effects this subsistence shift will have on our biology and social and cultural systems.

Despite the rising concern, however, about industrial food systems and overnutrition resulting in obesity, we must not lose focus on the millions of people who suffer from hunger and undernutrition. This is vividly illustrated in Casiday and colleagues' chapter on food shortages and acute malnutrition in Niger (chapter 7) and in Moffat and Finnis' (chapter 6) discussion of chronic malnutrition in periurban Nepal. However, under- and overnutrition must be examined together as problems emanating from a global food system. Indeed, malnutrition in both the forms of under and overnutrition can coexist in one nation and even within one individual's life course (Popkin 2001; Prentice 2005). Although in some parts of the world we have food surpluses—for example, the overproduction of corn grown in North America due to government subsidies (Pollan 2006)—we still have shortages of food in other parts of the world due to failures in agricultural production as well as to dysfunctional distribution. These are problems of politics and economic inequities that cannot be examined in isolation.

A new and growing challenge of the twenty-first century will be the rapidly changing agricultural conditions due to global climate change. Some researchers have begun to predict and model the effects of climate change in major agricultural areas such as India (Sinha and Swaminathan 1991) and Africa (Downing 1991). Finnis (2007) has described adaptive, though potentially unsustainable, responses on the part of small farmers in the Kolli Hills of Tamil Nadu, India who have replaced millet with cassava cash-crops due to recent erratic and unreliable rainfall patterns. More biocultural case studies of current social-cultural and ultimately biological responses to climate change will enable us to contribute to solving future climate-induced food crises.

Conclusion

Ultimately, our diet today is as fundamental to who we are and how we are evolving as it was for our ancestors. We have attempted to show throughout this book that regardless of the methods used, scale of analysis, and time frame, the biocultural approach to the study of human diet and nutrition provides a rich analytical framework for research. So too, a view of human diet and nutrition that encompasses a full span of evolutionary time (including the present) and a life course perspective allows us to gain a more holistic perspective on the human condition.

References

Aiello, L. C., and P. Wheeler. 1995. The expensive-tissue hypothesis: The brain and the digestive system in human and primate evolution. *Current Anthropology* 36 (2): 199–221.

Arnas, Y. A. 2006. The effects of television food advertising on children's food purchasing requests. *Pediatrics International* 48: 138–45.

Beaton, G. H. 1989. Small but healthy? Are we asking the right question? *Human Organization* 48 (1): 31–37.

Birchall, J., T. C. O'Connell, T. H. E. Heaton, and R. E. M. Hedges. 2005. Hydrogen isotope ratios in animal body protein reflect trophic level. *Journal of Animal Ecology* 74: 877–881.

Bird, D. W., and R. Bliege Bird. 2000. The ethnoarchaeology of juvenile foragers: shellfishing strategies among Meriam children. *Journal of Archaeological Anthropology* 19: 461–76.

Cachel, S. 1997. Dietary shifts and the European upper Palaeolithic transition. *Current Anthropology* 38 (4): 579–603.

Conklin-Brittain, N. L., R. W. Wrangham, and K. D. Hunt. 1998. Dietary response of chimpanzees and Cercopithecines to seasonal variation in fruit abundance: II. Macronutrients. *International Journal of Primatology* 19: 949–70.

Corr, L. L., J. C. Sealy, M. C. Horton, and M. P. Evershed. 2005. A novel marine dietary indicator utilizing compound-specific bone collagen amino acid $\delta^{13}C$ values of ancient humans. *Journal of Archaeological Science* 32 (3): 321–30.

Dirks, W., and J. E. Bowman. 2007. Life history theory and dental development in four species of catarrhine primates. *Journal of Human Evolution* 53: 309–20.

Downing, T. E. 1991. Vulnerability to hunger in Africa: A climate change perspective. *Global Environmental Change* December: 365–80.

Eaton, S. B., and M. Konner. 1985. Paleolithic nutrition: A consideration of its nature and current implications. *New England Journal of Medicine* 312 (5): 283–89.

Finnis, E. 2007. The political ecology of dietary transitions: Changing production and consumption patterns in the Kolli Hills, India. *Agriculture and Human Values* 24: 343–53.

Fogel, M. L., and N. Tuross. 2003. Extending the limits of palaeodietary studies of humans with compound specific carbon isotope analysis of amino acids. *Journal of Archaeological Science* 30 (5): 535–45.

Furuichi, T., C. Hashimoto, and Y. Tashiro. 2001. Fruit availability and habitat use by chimpanzees in the Kalinzu Forest, Uganda: Examination of fallback foods. *International Journal of Primatology* 22 (6): 929–45.

Goodenough, W. H. 2004. Anthropology in the 20th century and beyond. *American Anthropologist* 104 (2): 423–40.

Goodman, A. H., and G. J. Armelagos. 2000. Disease and death at Dr. Dickson's Mounds. In *Nutritional Anthropology. Biocultural Perspectives on Food and Nutrition,* ed. A. H. Goodman, D. L. Dufour, and G. H. Pelto, 58-61. Mountain View, CA: Mayfield Publishing Company.

Harlow, M., and R. Laurence. 2002 *Growing Up and Growing Old in Ancient Rome: A Life Course Approach.* London: Routledge.

Hawkes, K., J. F. O'Connell, and G. Blurton. 1995. Hadza children's foraging: Juvenile dependency, social arrangements, and dependency among mobile hunter-gatherers. *Current Anthropology* 36: 1–24.

Kaplan, H., K. Hill, J. Lancaster, and M. Hurtado. 2000. A theory of human life history evolution: Diet, intelligence and longevity. *Evolutionary Anthropology* 9 (4): 156–85.

Kuzawa, C. 2005. Fetal origins of developmental plasticity: Are fetal cues reliable predictors of future nutrition? *American Journal of Human Biology* 17: 5–21.

Laden, G., and R. Wrangham. 2005. The rise of hominids as an adaptive shift in fallback foods: Plant underground storage organs (USOs) and australopith origins. *Journal of Human Evolution* 49: 482–98.

Lambert, J. E., C. A. Chapman, R. W. Wrangham, and N. L. Conklin-Britttain. 2004. Hardness of cercopithecine foods: Implications for the critical function of enamel thickness in exploiting fallback foods. *American Journal of Physical Anthropology* 125: 363–98.

Landau, M. 1991. *Narratives of Human Evolution.* New Haven, CT: Yale University Press.

Leonard, W. R. 1994. Evolutionary perspectives on human nutrition: The influence of brain and body size on diet and metabolism. *American Journal of Human Biology* 6 (1): 77–88.

———. 2002. Food for thought. *Scientific American* 287 (6): 106–16.

Lewis, M. E. 2007. *The Bioarchaeology of Children.* Cambridge: Cambridge University Press.

Marshall, A. J., and R. W. Wrangham. 2007. Evolutionary consequences of fallback foods. *International Journal of Primatology* 28: 1219–35.

Martorell, R. 1989. Body size, adaptation and function. *Human Organization* 48 (1): 15–20.

McCullagh, J. S. O., D. Juchelka, and R. E. M. Hedges. 2006. Analysis of amino acid C-13 abundance from human and faunal bone collagen using liquid chromatography/isotope ratio mass spectrometry. *Rapid Communications in Mass Spectrometry* 20 (18): 2761–68.

McElroy, A. 1990. Biocultural models in studies of human health and adaptation. *Medical Anthropology Quarterly* 4: 243–65.

Messer, E. 1989. Small but healthy? Some cultural considerations. *Human Organization* 48 (1): 39–51.

Molleson, T., K. Jones, and S. Jones. 1993. Dietary changes and the effects of food preparation on microwear patterns in the Late Neolithic of Abu Hureyra, northern Syria. *Journal of Human Evolution* 24: 455–68.

Nestle, M. 2000. Paleolithic diets: A skeptical view. *Nutrition Bulletin* 25: 43–47.

O'Keefe, Jr., J. H., and L. Cordain. 2004. Cardiovascular disease resulting from a diet and lifestyle at odds with our Paleolithic genome: How to become a 21st-century hunter-gatherer. *Mayo Clinic Proceedings* 79 (1): 101–8.

Panter-Brick, C. 1998. *Biosocial Perspectives on Children.* New York: Cambridge University Press.

Panter-Brick, C., A. Todd, and R. Baker. 1996. Growth status of homeless Nepali boys: Do they differ from rural and urban controls? *Social Science and Medicine* 43: 441–51.

Pennisi, E. 1999. Did cooked tubers spur the evolution of big brains? *Science* 283: 2004–5.

Pollan, M. 2006. *The Omnivore's Dilemma: A Natural History of Four Meals.* New York: Penguin Books.

Popkin, B. M. 2001. The nutrition transition and obesity in the developing world. *Journal of Nutrition* 131: 871S–873S.

Prentice, A. M. 2005. The emerging epidemic of obesity in developing countries. *International Journal of Epidemiology* 35(1): 93–99.

Ritenbaugh, C. 1991. Body size and shape: A dialogue of culture and biology. *Medical Anthropology* 13: 173–80.

Schmidt, C. W. 2001. Dental microwear evidence for a dietary shift between two nonmaize-reliant prehistoric human populations from Indiana. *American Journal of Physical Anthropology* 114: 139–45.

Schwartz, G. T., P. Mahoney, P., Godfrey, L.R., Cuozzo, F., et al. 2005. Dental development in *Megaladpis edwardsi* (Primates, Lemuriformes): Implications for understanding life history variation in subfossil lemurs. *Journal of Human Evolution* 49: 702–21.

Schwartz, G. T., K. E. Samonds, L. R. Godfrey, W. L. Jungers et al. 2002. Dental microstructure and life history in subfossil Malagasy lemurs. *Proceedings of the National Academy of Science* 99: 6124–29.

Seckler, D. 1980. Malnutrition: An intellectual odyssey. *Western Journal of Agricultural Economics* 5 (2): 219–27.

Sinha, S. K., and M. S. Swaminathan. 1991. Deforestation, climate change, and sustainable nutrition security: A case study of India. *Climatic Change* 19: 201–9.

Ungar, P. 2004. Dental topography and diets of *Australopithecus afarensis* and early *Homo*. *Journal of Human Evolution* 46: 605–22.

Ward, C. 2003. The evolution of human origins. *American Anthropologist* 105 (1): 77–88.

White, F. J. 1998. Seasonality and socioecology: The importance of variation in fruit abundance to bonobo sociality. *International Journal of Primatology* 19: 1013–27.

Wiber, M. G. 1998. *Erect Men/Undulating Women: The Visual Imagery of Gender, "Race" and Progress in Reconstructive Illustrations of Human Evolution*. Waterloo, ON: Wilfrid Laurier University Press.

Wrangham, R. 2009. *Catching Fire: How Cooking Made Us Human.* New York: Basic Books.

Yamakoshi, G. 1998. Dietary responses to fruit scarcity of wild chimpanzees at Bossou, Guinea: Possible implications for ecological importance of tool use. *American Journal of Physical Anthropology* 106: 283–95.

Notes on Contributors

● ● ●

Sabrina C. Agarwal received her MSc and PhD (2001) from the Department of Anthropology, University of Toronto, and completed a two year Social Sciences and Humanities Research Council of Canada Postdoctoral Fellowship in the Department of Anthropology, McMaster University. She is currently Assistant Professor of Anthropology and faculty associate of the Archaeological Research Facility, University of California, Berkeley. Her research interests are broadly focused on the study of bone biology and health, with particular study of the role of aging, sex, and gender on bone remodeling and fragility. Her work has examined patterns of cortical bone microstructure, trabecular architecture, mineral density, and biomechanical properties in several archaeological populations in the Old World including the Anatolian Neolithic, Medieval and Post-Medieval Britain, and Imperial Rome, and she is currently also interested in the examination of bone maintenance in the nonhuman primate (monkey) model. Her publications include a coedited volume, *Bone Loss and Osteoporosis: An Anthropological Perspective,* 2003 (Kluwer Academic/Plenum) with S. D. Stout.

Luca Bondioli heads the Section of Anthropology of the Museo Nazionale Preistorico Etnografico "Luigi Pigorini" in Rome and is the curator of the osteological collections of the same museum. The Section of Anthropology's main research activities are in paleobiology of skeletal populations with special reference to paleonutrition, paleopathology, dental anthropology, adaptation, and variability. For many years the Section has focused on the study of the locomotory patterns in fossil and extant primates including *Homo* and on the Early Middle Pleistocene human peopling in the Horn of Africa. The Section develops advanced research methods with special reference to microtomography, digital image processing, and data visualization.

Erin Bryce is currently a doctoral candidate in the Department of Anthropology, University of Toronto. Her research interests include biological demography and the effects of weather and climate on human health.

Stacie Burke is an Associate Professor in the Department of Anthropology, University of Manitoba. Current research interests are focused in the area of colonialism and health, the sanatorium era of tuberculosis treatment in Ontario, and interwar transitions in the practice of medicine in Canada. Recent publications can be found in the *Canadian Bulletin of Medical History* and *Medical History*.

Rachel Casiday completed her PhD in medical anthropology at Durham University in 2005 and currently lectures in the Department of Voluntary Sector Studies at the University of Wales, Lampeter. She has research interests, broadly speaking, in the social context of risk, children's health, the health impacts of poverty and inequality, and voluntary and social action to address health problems.

Tosha L. Dupras is an Associate Professor of Anthropology at the University of Central Florida. Her specializations include human osteology, stable isotope analysis, human growth and development, Egypt, bioarchaeology, and forensic archaeology. She has been associated with two archaeological expeditions in Egypt—the Dakhleh Oasis Project since 1995 and the Dayr al-Barsha Expedition since 2004—where she excavates and analyses skeletal material. Her publications can be found in journals such as the *American Journal of Physical Anthropology, Journal of Archaeological Sciences,* and the *Journal of Osteoarchaeology*. She is coauthor of two books, *Osteology of Infants and Children* (Texas A&M) and *Forensic Recovery of Human Remains: Archaeological Approaches* (CRC Press).

Elizabeth Finnis is an Assistant Professor in the Department of Sociology and Anthropology at the University of Guelph. Her major research interests include the political ecology of food, diet, and agricultural transitions, and environment-economics tensions in the context of agricultural practices among small farmer households in South India and Paraguay. Her work has been published in a variety of journals, including *American Anthropologist, Agriculture and Human Values,* and *Food, Culture and Society*.

Charles Fitzgerald received his PhD from the University of Cambridge in 1996, and since then his research has been mainly focused on utilizing histological microstructures of teeth to explore issues in dental development of anthropological interest. After completing postdoctoral fellowships at McMaster University and then at University College London, he worked with Shelley Saunders at the McMaster University Anthropology Hard Tissue and Light Microscopy Laboratory from 2003 to Shelley Saunders' death in 2008. His ongoing research is the analysis of deciduous tooth germs to determine precise age of death of infants from the Kylindra cemetery on the Greek Island of Astypalaia. This project is in collaboration with Simon Hillson from UCL and the 22nd Ephorate of Prehistoric & Classical Antiquities.

Bonnie Glencross is an Assistant Professor in the Department of Archaeology and Classical Studies at Wilfrid Laurier University in Ontario, Canada. Bonnie studied at the University of Toronto where she received her BSc, MA and PhD (2003) in Anthropology, and produced her doctoral thesis on skeletal injury patterns and lifetime fracture risk in prehistoric hunter-gatherers from Indian Knoll, Kentucky. From 2006–08 she held a Social Sciences and Humanities Research Council (SSHRC) Postdoctoral Fellowship in the Department of Anthropology, University of California Berkeley. Her postdoctoral research remains a focus and is contributing to a long term collaborative and interdisciplinary investigation of biocultural adaptations in the Neolithic community of Çatalhöyük, Turkey. In addition to her research in Turkey, she is also the primary investigator for a project on work-related injury and fatalities during the Victorian era in Ontario. She has published in IJO on the identification of childhood skeletal trauma in the archaeological record and most recently co-edited and contributed to the volume Social Bioarchaeology, Global Archaeology Series, Wiley-Blackwell Press (in press).

Kate R. Hampshire is a Senior Lecturer in Anthropology at Durham University, where she has worked since completing her PhD (at University College London) in 1998. Her primary research interests are in child health and spatial mobility in sub-Saharan Africa. Her recent work in Africa includes projects on children's daily mobility in Ghana, Malawi, and South Africa (ESRC/DFID-funded) and on social support among young Liberian refugees in Ghana (Nuffield funded), as well as the work in Niger on which her essay in this volume is based. She also has several current UK-based research projects, including one on experiences of infertility among British Pakistanis (ESRC funded) and another looking at the impacts of arts programmes on adolescents' well-being (Arts Council funded).

Lesley Harrington received her PhD from the University of Toronto and is currently Social Sciences and Humanities Research Council of Canada Postdoctoral Fellow in the Department of Palaeontology at The Natural History Museum in London. Her doctoral research focused on reconstructing the physical activities of prehistoric hunter-gatherer children as reflected in the development of postcranial bone mass. Her broader research interests include modern human variation in growth and development, and prehistoric breastfeeding and weaning behavior.

Peter Katzmarzyk is currently a Professor and the Associate Executive Director for Population Science at the Pennington Biomedical Research Center in Baton Rouge, Louisiana, USA. He also holds the Louisiana Public Facilities Authority Endowed Chair in Nutrition. He obtained a PhD in Exercise Science from Michigan State University in 1997, and pursued post-doctoral education at Laval University in 1998. His main research interest is the epidemiology and public health impact of obesity and physical inactivity, and determining the relationships between physical activity, physical fitness, obesity and related disorders such as metabolic syndrome, cardiovascular disease and diabetes. He has published his research findings in more than 190 scholarly journals and books, and regularly participates in the scientific meetings of several national and international organizations. He is currently an editorial board member for the International Journal of Pediatric Obesity, Journal of Physical Activity and Health, and Metabolic Syndrome and Related Disorders.

Kate Kilpatrick worked as a program coordinator with Concern Worldwide's emergency nutrition program in Niger during 2005–6. She is currently based in Bonn, Germany, where she works as a consultant researcher and project manager for development and humanitarian organizations. Her areas of particular interest and expertise include livelihoods, health, and the environment.

William R. Leonard is Professor and Chair of Anthropology and Director of Global Health Studies Program at Northwestern University. His research examines how social and ecological stressors influence human biological variation and health in contemporary and prehistoric populations. He has published numerous refereed articles and coedited the volume *Human Biology of Pastoral Populations* (2002).

Roberto Macchiarelli is a professor of human evolution at the National Museum of Natural History (MNHN) of Paris and of paleobiology and evolution at the Department of Geosciences of the University of Poitiers, where he also coordinates the interdepartmental Center of Microtomography (UdP-CdM). A member of the French CNRS research laboratory at the MNHN Dept. of Prehistory (UMR 7194) and president of the NESPOS Society (*Neanderthal Studies Professional Online Service*), he worked for twenty-three years at the Laboratory of Palaeobiology of the National Archaeological Service of Abruzzi and at the Section of Anthropology of the National Prehistoric Museum of Rome. He is author of over 150 scientific papers and his major research interests include human odontoskeletal paleobiology, evolutionary anatomy, and functional morphology. Founder member of the European Anthropological Association, he is coeditor of the *Digital Archives of Human Paleobiology*. Since 2004, he directs the PaleoY research project in Yemen.

Tina Moffat is an Associate Professor in the Department of Anthropology at McMaster University, Canada. Her research focuses on child health and nutrition in relation to environmental health and urban ecosystems. She grounds her research in biocultural and political-economic approaches, with main geographic areas of focus in Nepal and Canada. She has authored and coauthored numerous scholarly journal publications on child growth and infant feeding in Nepal, and nutritional well-being and obesity among school children in Canada.

Catherine Panter-Brick is a Professor of Anthropology, Health, and Global Affairs in the Department of Anthropology at Yale University. Her research focuses on critical risks to health across key stages of human development. She has edited several books to bridge research findings into teaching practice, such as *Biosocial Perspectives on Children* (1998), *Hormones, Health, and Behavior* (1999), *Abandoned Children* (2000), *Hunter-Gatherers* (2001), and *Health, Risk, and Adversity* (2009). She is Senior Editor (Medical Anthropology Section) for *Social Science & Medicine*.

Susan Pfeiffer is a Professor in the Department of Anthropology, University of Toronto. Her research in biological anthropology focuses on the reconstruction of past human adaptations through analysis of the human skeleton. Particular interests include analyses of past hunter-gatherers of Southern Africa, past populations of the North American Great Lakes region, and methodological work on estimation of age at death from human cortical bone tissue. Her research program emphasizes life course characteristics like growth, age at death, nutritional reconstruction, and health indicators. She is an honorary research associate of the Department of Archaeology, University of Cape Town. She has published widely.

Tracy L. Prowse was previously an Assistant Professor of Anthropology at Southern Illinois University and is currently an Assistant Professor at McMaster University. Her research explores diet and health in past populations using paleopathological and isotopic analyses of human bones and teeth. She investigates the social, historical, and political economic conditions to examine their impact on diet, health, and mobility of people living in the Mediterranean region. She has published on the paleodiet of Roman Italy in the *American Journal of Physical Anthropology* and the *Journal of Archaeological Science*.

Marcia L Robertson is a research associate in the Department of Anthropology at Northwestern University. Her work has examined the ecology of early hominids and miocene hominoids. She has published numerous articles on the evolution of locomotor strategies and nutritional needs in early hominids.

Shelley R. Saunders, tragically, died shortly before the publication of this volume. Over her working life she made contributions to many areas of biological anthropology and had an enormous impact on the discipline, particularly in Canada. She received her PhD in anthropology from the University of Toronto in 1977 and spent most of her career as a professor in the Department of Anthropology at McMaster University. She held a Canada Research Chair in Human Disease and Population Origins and was the founder of the McMaster Ancient DNA Centre and the McMaster Anthropology Hard Tissue and Light Microscopy Laboratory. In 2001 she became the first anthropologist to be elected to the Royal Society of Canada. Professor

Saunders was a superb teacher and many of her students have gone on to positions in anthropology at universities in Canada and throughout the world.

Larry Sawchuk received his PhD from the University of Toronto. He is an Associate Professor at University of Toronto who specializes in medical and demographic anthropology. He is interested in the health of colonial populations, with research focal points primarily in Gibraltar and Malta.

Daniel Sellen conducts research on the human ecology, evolutionary biology, and global health consequences of young child feeding and care-giving practices. He is Canada Research Chair in Human Ecology and Public Nutrition and Associate Professor in three Departments (Anthropology, Nutritional Sciences, and Public Health Sciences) at the University of Toronto. Professor Sellen was trained in zoology at Oxford (BA, MA, 1987), anthropology at Michigan (MA 1989), and theoretical ecology and international nutrition at the University of California, Davis (PhD 1995). He then worked at University College London (as a Leverhulme Trust Postdoctoral Fellow in Demographic Anthropology) and at the London School of Hygiene and Tropical Medicine (as both a Visiting and Honorary Lecturer in Public Health Nutrition). He taught previously at Emory University, where he maintains an affiliation as Adjunct Associate Professor of Anthropology and Global Health. He has served on numerous editorial boards, review panels, and professional committees.

Josh Snodgrass is an Assistant Professor in the Department of Anthropology at the University of Oregon. He has affiliations with the Center for Ecology and Evolutionary Biology and the Institute of Cognitive and Decision Sciences at the University of Oregon. Dr. Snodgrass received his MA in Anthropology from the University of Florida and his PhD in Anthropology from Northwestern University. Following a National Institute on Aging postdoctoral fellowship at the University of Chicago, he moved to the University of Oregon. His research focuses on human adaptation to environmental stressors, the influence of economic development on health and nutritional status, and the evolution of human and primate energy requirements.

Glossary

Acute malnutrition: refers to wasting (thinness) and/or nutritional edema. Acute malnutrition is present when an individual is considered too thin for their particular height.
Adipose tissue: loose connective tissue that holds fat cells
Adiposity: the state of being fat
Abdominal obesity: excess fat deposits in the abdominal region
Adze: a tool resembling a hoe that is used to smooth rough-cut wood in hand woodworking
Allometric: denoting the change of proportion between organs or parts during the growth of an organism
Ameloblasts: cells that form enamel
Amphora(e): a ceramic vessel usually with a long neck and two handles; used in antiquity for transportation of liquids (e.g., wine, oil)
Anthropometry: the study of the size and proportions of the human body
Anthropometric data: information resulting from the scientific study of measurements of the human body
Apomorphies: derived characters that are not shared with an organism's ancestors and therefore likely to be specialized and recent adaptations
Australopithecus: the genus reserved for extinct members of the family Hominidae who have relatively small brains, a reduction in tooth size when compared to earlier primates, and were capable of at least partial bipedalism
Basiocciput: the part of the occipital bone that lies anterior to the foramen magnum and joins with the body of the sphenoid bone
Biocultural Approach: a holistic perspective that includes the use of evolutionary, ecological, cultural, and political-economic frameworks to investigate human biology in social context. "Biosocial" is the preferred term in the United Kingdom.
Body Mass Index (BMI): a measure of an individual's weight relative to their height, calculated as weight (kg) / [height (m)]^2

Brown striae of Retzius: regularly occurring growth layers visible in microscopic thin sections of teeth
Cappuccina burial: a burial covered with large tiles in the form of an inverted V
Chronic malnutrition: found among children who have inadequate height for age relative to a population growth reference
Chronometric dating: the determination of age with reference to a specific time scale, also known as *absolute dating*
Circadian (diurnal) rhythm: physiological regulation of certain bodily processes; occurs on an approximately twenty-four-hour cycle
Celiac disease: a disease of the digestive tract caused by an immunological reaction to gluten, a protein found in wheat and other grains
Cohort: a group of individuals belonging to the same age category
Collagen: a structural protein found in bone
Colostrum: a fluid secreted by the breasts around the time of birth (prior to the arrival of breast milk); a source of nutrients and antibodies for the infant
Complementary foods: nutritionally rich and relatively sterile combinations of foods acquired and processed by caregivers and fed to breastfed infants and toddlers after about six months of age
Cribra orbitalia: pathological lesions at the roof of the eye orbits characterized by porosity, pitting, or new bone growth. It is typically associated with iron deficiency in children.
Cross-sectional surveys: a research method where data are collected from a subset or subsets of a population at a single point in time
Cystic fibrosis: a hereditary disease that affects the movement of salt and water in the body's cells; causes the lungs and pancreas to secrete thick mucus
Day range: the distance that an animal travels (in km) in a twenty-four-hour period.
Diet Quality Index (DQ): Reflects the relative proportions (percentage by volume) of (1) structural plant parts (e.g., leaves, stems, bark), (2) reproductive plant parts (e.g., fruits, flowers), and (3) animal foods (including invertebrates). The index ranges from a minimum of 100 (a diet of all leaves and/or structural plant parts) to 350 (a diet of all animal material).
Dietary delocalization: when an increasing proportion of the daily diet comes from foreign sources, usually through commercial channels
Dietary Diversity Score (DDS): a score determined by the number of different food groups consumed over a given time
Dakhleh Oasis Project (DOP): a long-term regional study of the interaction between environmental changes and human activity in the closed area of the Dakhleh Oasis, Western Desert of Egypt
Doubly labeled water method: uses the naturally occurring stable isotopes of water (D_2O and $H_2^{18}O$) to assess energy expenditure, body composition, and water flux in humans and animals

Emic: a research approach characterized by attempting to understanding meaning from the group member's point of view

Encephalization: in the context of human evolution, it is the tendency towards larger brain size in hominids throughout our evolutionary history

Etic: a research approach characterized by analytical, outside perspectives regarding the interpretation of meaning

Epidemiology: the branch of medicine dealing with the amount of disease in populations and with detection of the source and cause of epidemics of infectious disease

Exclusive suckling: a life history phase during which a juvenile mammal derives all nutrients from maternal milk. It is often referred to as "infancy" in nonhuman mammals.

Exogenous: developed or originating outside of the organism or local environment

Enamel dentine junction (EDJ): the boundary between the enamel and dentine on a tooth

Family foods: raw foods and combinations of foods collected, processed, and shared by older juveniles and adults and consumed by older members of the family

FAO: Food and Agriculture Organization of the United Nations

Febrile disease: any disease characterized by a high fever

Food commoditization: the use of agricultural goods for sale rather than for home consumption

Food Variety Score (FVS): a score generated by the mean number of different food items consumed from all possible items eaten

Foraging: the collection of wild plants and the hunting of wild animals for subsistence

Fynbos: low-growing and evergreen vegetation found mostly in the Western Cape of South Africa

Gestation: period of time between fertilization and birth; commonly called *pregnancy*

Global acute malnutrition (GAM): a measure of malnutrition where a child's weight-for-height z-score is < −2 standard deviations below the reference median of the population growth reference

Growth failure/faltering: growth rate below the appropriate growth velocity for a given age relative to a population growth reference

Growth stunting: defined as height for age below the fifth percentile (or -2 standard deviations below the median) relative to a population growth reference; used as an indicator of chronic malnutrition in children

Herbaceous plants: perennial, nonwoody plants that lie dormant in winter and produce new growth in the spring

Herbivore: an organism that consumes predominantly plants

Home range: the total area (in hectares) used by a group of animals; usually composed of more than one day range

Hominid: the family Hominidae, which includes the genera *Australopithecus* and *Homo*

Homo: the genus reserved for individuals in the human lineage of which *Homo sapiens* are the only extant species

Iliac crest: superior border on the blade of the ilium

Infancy: in humans, the period between birth and the first birthday; in other mammals, the period of dependency on mother's milk

Kleiber Relationship: describes the correlation between body weight and resting metabolic rate (RMR). Metabolic rate increases as a function of body weight raised to the 3/4th power.

Lactation: the ability to secrete immunologically active and nutritious milk from ventral epidermal glands

Libation burial: a burial with a broken amphora or tile tube visible on the cemetery surface used to pour liquid libations to the deceased

Linear enamel hypoplasia (LEH): a line, groove, or pit normally visible on the surface of a tooth; caused by developmental disturbances during enamel matrix formation

Macronutrients: energy-yielding nutrients such as proteins, fats, and carbohydrates

Microlithic: tools made of small stones, typically knapped of chert or flint

Micronutrients: vitamins and minerals; regulators that assist in all body processes

Moral economy: the interplay between moral or cultural beliefs and economic activities

Morbidity: relating to the incidence of disease in a population

Mortality: relating to the number of deaths in a population

Multiparous: a woman who has given birth to multiple children

Necropolis: a cemetery, often associated with an ancient city

Neonatal line: a pronounced Wilson band that is present in any tooth that is forming prior to birth; it is formed around the time of birth

Neonate: an infant aged one month or less

Obesity: a condition where there is excess body weight due to an abnormal accumulation of fat. Obesity is determined by a Body Mass Index of 30 kg/m^2, or more.

Odontochronology: the assignment of chronology to dental development events based on interpreting incremental microstructures in teeth

Overweight: Body Mass Index (BMI) between 25–29.9 kg/m^2

Obstetric canal: the birth canal of a female, consisting of the cervix, vagina, and vulva

Overnutrition: a form of malnutrition in which nutrients are oversupplied relative to the amounts required for normal growth, development, and metabolism

Pap: a soft food for infants usually made from bread mixed with water or milk

Passive immunity: certain antibodies transferred from the mother to the fetus during pregnancy and to the infant while breastfeeding

Perinate: a fetus or infant between 27 days after birth and 20 weeks gestation
Periurban: a region that spans the landscape between contiguous urban development and rural countryside, has low population density, and encompasses a mix of land uses
Physical activity level (PAL): a ratio of Total Daily Energy Expenditure (TDEE) to Resting Metabolic Rate (RMR)
Pit burial: a simple burial with no associated burial structure
Plesiomorphies: shared primitive characters no older than the last common ancestor of a phylogenetic group of organisms
Polychlorinated biphenyls (PCBs): widely used, humanmade chemicals that have been demonstrated to cause cancer and other adverse health conditions
Physiological fractionation: variation in the isotope ratios as a result of physiological processes in the body, as well as a function of their atomic mass
Porotic hyperostosis: a condition characterized by the overgrowth of the spongy marrow space of the skull. Typically results in porosity and skeletal lesions at the outer table of the cranial vault. This condition is associated with iron deficiency resulting from dietary inadequacies or parasite infection.
Postpartum: the period following the birth of an offspring
Preschoolers: in humans, children between the ages of one and five years
Primiparous: a woman who has given birth to only one child
Qualitative: referring to research methods that involve nonmetric data
Quantitative: referring to research methods that involve metric data
Resting metabolic rate (RMR): the energy required by the human body at rest to maintain basic physiological functions (measured in kcal/day)
Rickets: a softening of the bones in children leading to characteristic skeletal deformities, one of which is "bowed legs" or "knocked knees"
Seasonality: the changing availability of resources according to different seasons of the year
Secular change: Change occurring over a long period of time or across generations
Socioeconomic status (SES): a measure of an individual's place within a social group based on various factors, including occupation, education, income, wealth, and place of residence
Severe acute malnutrition (SAM): a measure of malnutrition where the child's weight-for-height z-score is < -3 standard deviations below the median of the population growth reference
Subsistence farming: farming that provides for the basic needs of the farmer and their family without surpluses for marketing
Suckling: the mechanical process of extracting milk from the mammary glands and/or ducts. In humans, suckling is a component of "nursing" or "breastfeeding" by a mother or baby. It is interesting that in English and many other languages such verbs carry both active and passive meanings.

Sudden infant death syndrome (SIDS): sudden death of an infant from unexplained causes; often occurs during sleep

Swaddling: the process of wrapping a baby in bands of cloth

Synapomorphies: evolutionarily derived or specialized characters shared only by one phylogenetic group of organisms

Total daily energy expenditure (TDEE): total amount of energy used in all activities throughout a twenty-four-hour period (measured in kcal/day). This can also be estimated using doubly labeled water or heart rate monitoring.

Transitional feeding: a life history phase during which nutrition is derived from a combination of maternal milk and other foods foraged by the infant, its parents, or others. It is poorly described for most primates.

Trophic (level): pertaining to nutrition or to a position in a food chain, food web, or food pyramid

Undernutrition: malnutrition due to inadequate food supply or to inability to metabolize or use necessary food elements

Urbanization: the increase over time in a population of cities in relation to the region's rural population

Weaning: the termination of suckling

Weaning process: the process by which weaning occurs in humans with a gradual reduction of breastmilk and the introduction of complementary foods until the final termination of breastfeeding

Weanling: a life history phase during which a recently weaned juvenile mammal must forage for itself and subsist on foods similar or identical to those selected by adults

Wet nurse: a woman who is hired to breast feed an infant who is not her own child

Wilson bands (accentuated striae of Retzius; pathological striae): Microscopic defect in the enamel caused by developmental disturbances during enamel matrix formation

Young childhood: in humans, the period between the first and third birthday

Index

Adaptation, 37, 58–59, 173, 242
  adaptive shift, 19, 245
  lactation and, 59–64, 71–75
  morphological, 51–52, 243
Adiposity, 22, 29, 59, 223–28, 232, glossary
  maternal, 63, 66, 70, 71, 74, 228
  See also Obesity
Adolescence, 40–42, 48, 52, 177, 180, 202–3, 209, 224–25
  growth spurt, 42, 51, 180
Africa, 19–20, 90, 92, 113, 208–10, 245, 248
  !Kung and Khoe-San, chapter 2
  Dakhleh, 5, 44–45, chapter 4
  Egypt, 5, 44–45, chapter 4, 176–77, 244
  Kalahari, 37, 39–40, 51
  Niger, 6, chapter 7, 244, 248
  South Africa, 4, chapter 2, 139
Age, 40–43, 48, 181
  chronological age, 42, 179, 181, 185
  skeletal and dental age estimates, 40–44, 48, 176, 179, 185
  See also Aging
Agriculture, 2, 6, 21, 50, 113, 134, 137, 153, 228, 244–48
  agriculturalists, 27, 206–7
  cash-cropping, 145, 248
  See also Farming; Food, food production
Aging, 64, 197, 200
  See also Age

Allometry, 41, 51
Anthropometrics, 144, 153, 224–225, glossary
  See also Growth and development
Asia, 21, 134, 137, 248
  Nepal, 6, chapter 6, 244, 248

Biocultural, Introduction, 59, 92, 102, 146, 174, 181, 198, 203, conclusion, glossary
Bipedalism, 3, 64, 71, 73
Birth (also Childbirth), 41, 51, 61, 63, 111, 113, 116, 123
  Birth interval, 4, 39, 59, 64, 65, 72, 246
  See also Growth and development, birth weight; Reproduction
Body fat. *See* Adiposity; Obesity
Body Mass Index (BMI), 224–33
  calculation of, glossary
Body size, 4, 14–21, 37–42, 60, 64, 243
Bone, 3, 5, 7, 37, 90–91, 120, 181–82, chapter 9, 247
  bone mineral density (BMD), chapter 9
  cells, 199, 202, 214
  formation, 187–88, chapter 9
  growth, 40–42, 51, 180–82, 185–87, chapter 9
  loss. *See* Osteoporosis
  mass, 7, 38, chapter 9
  pathology, 212
  See also Skeleton

Brain size or growth, 4, 14–17, 22, 29, 61
  hominid, 3, 18–21, 64, 71, 73, 228, 242, 246
Breastfeeding, 39, 45, chapters 3–5, 137–40, 155, 162–65, 176–77, 181–82, 186, 188, 210, 242–47
  breast milk, 39, 58, 63–73, 91–92, 96, 99, 123, 139–40, 144, 162–63, 186
  colostrum, 73, 176, 187
  secular change/trends in, 5, 109, 112, 116–18, 126
  wet nursing, 176–77, glossary
  See also Infant feeding; Lactation; Weaning
Burial, 174

Caloric intake, 6, 14–30, 35, 60–63, 70, 139, 145, 223, 228, 242, 246
  See also Diet
Carnivore, 20, 91
Centers for Disease Control and Prevention (CDC), 225
Childhood, 5, 35, 38, 40, 52, 58, 167, 173, 246
  child morbidity and mortality, 50, 72, 98, 155–6, 161, 168, 173, 181, 184–88, glossary
  childcare, 68, 75, 102, 118–23, 164, 173
  diet and nutrition, 6–7, 22–41, 123, chapters 6–7, 177, 243, 246–47
  See also Growth and development; Health, child health; IYCF
Circadian rhythm, 179, glossary
Climate, 137, 248
Colonialism, 6, 113, 120–21
Concern Worldwide, 154, 159–61, 168

Demography, 59, 72, 89, 92, 94, 102
Dentition, 3, 5, 19, 64, 80, 90–92, 97–101, 174, 176, 178, 188, 245
  development of, 40–44, 83, 92, 99–101, 176, 178–80, 185, 189
  linear enamel hypoplasia (LEH), 89, 178–79, glossary
  neonatal line, 179–80, glossary
  microstructure, 177–83, 187, 189

  See also Age, dental and skeletal age estimates
Development (economic)
  developing world, 5–6, 14, 22–24, 110
  industrialization, 14, 24–31, 68, 72, 74, 118, 121, 124, 126, 134, 245–48
Diabetes, 6, 110, 134, 232–35, 242
Diet
  dietary delocalization, 2, 6, 135, 145, glossary
  dietary diversity, 6, chapter 6, glossary
  dietary quality, 9, 16–17, 20, 23–24, 29, 61, 147, 246, glossary
  dietary transitions, 6, chapter 6, 244
  See also Food; Nutrition
Disease, 73–74, 98, 110, 162, 166, 168, 177, 182, 206
  anemia, 40, 120
  cancer, 6, 139, 232–35
  cardiovascular, 6, 110, 232
  epidemiology, 7, 57, 64, 68, 72, 112, 200, 202–3, 212–14, 224, 243, glossary
  gastrointestinal, 111, 138, 143, 157, 168, 173, 178
  infectious, 6, 23, 36, 39, 110, 143, 153, 155, 157, 168, 173, 178, 180, 182, 184, 188, 206
  See also Diabetes; Osteoporosis
Division of labour, 37, 115

Ecology, 4–5, 19, 57–64, 70–71
Ecosystems, 19–21, 36, 43
Emic/etic perspectives, 161, glossary
Energy, 19, 29, 63, 70, 228
  expenditure, 4, 15, 21, 25–30, 37, 40, 60, 62 ,70, 144, 228, 243
  in food. *See* Caloric Intake
  measurement, 26
  physiological requirements, 14–16, 19–30, 62–70
  See also Metabolism
Environment
  environmental changes, 19, 102
  environmental hazards, 35
Eocene, 60

Europe, 3, 44, 90, 110–14, 134, 197, 209–12, 244
  Gibraltar, 5, chapter 5, 222, 224
  Italy, 7, 139, chapter 8, 244
  Spain, 113–14, 122, 124, 126
  United Kingdom, 44, 92, 112–14, 122, 209–10
Evolution, 2–3, 7, 58–59, 65, 73, 246
  evolutionary features, 58–59, 64, 66
  human evolution, 2–4, chapter 1, 35–36, 39, 51–52, chapter 3, 134, 198, 227–28, 242–48
  natural selection, 4, 59, 228
  See also Hominins
Famine, 134, 153, 158
  See also Hunger
Farming, 21–22, 25, 30, 72, 134, 145, 147, 158, 206, 248
  subsistence farming, 69, 135, glossary
  See also Agriculture
Feeding, 4, 6, 16, 61–63, 146, 155–56, 159–60, 164–68
  See also IYCF
Food
  animal, 16–19, 23, 36, 38, 50, 71, 134, 139–45, 177, 242, 246
  cereals and grains, 90, 96–99, 110, 137–46, 153–54, 158–65, 177, 206–7, 248
  commercial, 134–35, 141–48, 247
  cooking, 21, 67, 71, 73, 242, 246
  dairy products, 98–99, 110, 120, 139–45, 167, 177
  fallback foods, 245
  fats, 24–25, 38, 66, 134, 143, 147–48, 154, 159–60, 165, 247
  fruits and vegetables, 18, 21, 90, 139, 144, 177, 242
  plant foods, 16, 19, 21, 90, 177
  processing, 21, 58, 71, 135, 181, 247
  production, 1–2, 6, 23, 134, 153, 206. *See also* Agriculture
  program, 6, 158, 164
  security, 2, 5–6, 35, 138–39, 144, 152–58, 161, 166, 168, 228, 244, 248
  sharing, 68, 71, 73, 166, 242

staple food, 16, 21, 36, 140
supplements, 155–68. *See also* Nutrition, emergency nutrition programs
underground storage organs (USOs), 36, 245
See also Diet
Food and Agriculture Organization of the United Nations, 25, 139, glossary
Foragers, 4, 16, 19–20, 26, chapter 2, 69, 72, 134, 245
Foraging, 2, 4, 14, 19–21, chapter 2, 58, 61–63, 67–69, 71, 245, 247, glossary

Gender, 37, 115
Growth and development, 3–7, 14, 21–24, 29, chapter 2, 59, 64–67, 70–71, 74, 92, 94, 109, 137–38, 143, 148, 155, chapter 8, 199, 202–4, 208–9, 224, 243
  birth weight, 39, 44, 51, 74, 139, 178, 202
  body mass/weight, 14, 18, 20, 22, 24, 27–28, 38–40, 71, 118, 120, 123, 159, 167, 223–27
  of the brain. *See* Brain growth
  catch-up growth, 47, 87
  failure, 4, 7, 14, 21–23, 29, 35, 40–42, 47, 51, 137–38, 148, 156–57, 173–74, 180, 185, 209, glossary
  growth spurt. *See* Adolescence
  measures of, 155–59, 165, 232
  stature, 22, 29, 37–44, 51, 180, 185
  tempo, 4, 23, chapter 2, 64, 180–81, 184–85, 202
  See also Anthropometrics

Health
  bone health, 198, 204, 206, 214
  child health, 35–36, 57, 72, 74, 109, 123, 138, 143, 155, 157, 160–62, 166, 168, 178, 185, 188, 224–25, 230–32, 236
  disease and, 6–7, 94
  healthcare, 136, 138, 161

health risk, 224–25, 230–32
infant, 5, 39, 41, 57, 72–74, 110–13, 118–21, 162, 176–77, 187–88
maternal, 70, 74, 110
nutritional, 2, 28, 121, 134–35, 147, 153, 243. *See also* nutrition
public health, 73, 110, 116, 118, 232–36
Health surveys, 24, 46–49, 228–35
Height-for-age
See Growth and development, measures of
Herbivore, 19–20, 91, glossary
Herders, 26, 69
Herding, 37, 72
See also pastoralism
Holocene, 4, 37–41, 51
Hominins, 2–4, 14, 18–21, 36, 60, 63, 65, 227–28, 242–46
fossil evidence, 14, 18, 227, 245–46
Hormones, 198–210, 247
Horticulture, 51
Hunting and gathering, 2, 19–20, 27, 35–38, 51, 64, 69, 206, 212, 228, 242, 245–47
See also Foraging

Immunity
auto-immune diseases, 6
breastfeeding and, 58–59, 66, 68, 79, 176, 188
immune system 59
Infancy, 5, 21–22, 29, 64, 68, 94–95, 123, 166, 174–78, 187, 202, glossary
infant health. *See* Health, infant
infant morbidity and mortality, 68, 72, 98, 102, 110, 114–15, 153, 178, 181–82, 186–88, glossary
infant nutrition, 24, 60–70, 110–11, 121, 139, 160, 162, 174, 246
physiological development, 61, 64, 74, 202, 246
See also Growth and development
Infant feeding, 5, 39, 71, 74, chapter 5, 153, 165, 176, 187, 244
artificial feeding, 73, 111–12, 116, 118, 121–24, 126

complementary foods, 39, 57–58, 66–75, 91, 96, 98, 101, 110, 120, 137, 165, 177–78, 181–82, 187–88, glossary
infant formula, 110–11, 118, 120–21, 139–41, 162
IYCF (infant and young child feeding), 5, 7, 47, 39, chapters 3–5, 140–41, 144, 151, 163, 173, 176–77, 181, 242–44, 246
transitional feeding, 4, 61–63, 66–67, 72–74, 90–91, 94, 99, 102–3, 177–78, 181, 188, 246, glossary
See also Breastfeeding; Weaning
Iron Age, 51

Juveniles, 64, 71, 198, 202, 246
feeding, 4, 58, 61–62, 64, 68
health, 174
mortality, 72, 174
See also Growth and development

Kleiber relationship, 14, 16, glossary

Lactation, 4, chapter 3, 197–98, 203–4, 208–10, 228, glossary
See also Breastfeeding; Primates, lactation biology
Life history
events, 4, 213. *See also* Birth
lactation and, chapter 3
theory, definition of, 4, 59, 246
variables, 61, 72, 246

Mammals, 4, 14–16, 19, 36, 38, 58–60, 64, 66, 69, 178, 228, 242
See also Primates
Maternal, 5, 40, 59, 63–65, 99, 111, 120–23, 138, 146, 166, 181, 201–3, 210
See also Breastfeeding; Health, maternal; Lactation
Metabolism, chapter 1, 59, 71, 198, 206–8, 211, 213
of the brain, 4, chapter 1
metabolic disorders, 13, 21, 24, 41, 211–12

metabolic requirements, 14, 29, 60, 63
Resting Metabolic Rate (RMR), 14–15, 21–22, 25, 27, 29, glossary
See also Energy
Middens, 36, 38
Mid-upper-arm circumference (MUAC)
See Growth and development, measures of
Migration, 6, 92, 94, 113, 134–37, 144–47
Milk
See Breastfeeding, breast milk; Food, dairy products; Infant feeding, artificial feeding; Primates, lactation biology

National Institutes of Health (NIH), 224
Neolithic, 206, 211
North America, 3, 7, 21, 47–51, 90, 110–11, 134, 139, 206, 212, chapter 10, 248
 Canada, 7, 27, 188, 225–35
 Mexico, 2, 7, 23, 145, 232–33
 United States, 7, 14, 24–25, 30, 46–51, 111–12, 158, 197, 206, 225–35
Nutrients, 2, 14–17, 29, 58–59, 63, 67–68, 71, 138–39, 144, 197–207
 macronutrients, 7, 24–25, 36–39, 60–61, 66, 90–91, 101, 134, 144, 147–48, 177, 197, 202–8, 243–44, glossary
 micronutrients, 6–7, 41, 62, 120, 139, 143–44, 185, 187, 197–98, 200–13, glossary
Nutrition
 emergency nutrition program, chapter 7
 See also Food, supplements
 malnutrition, 2, 5–6, 110, chapter 7, 206, 248
 nutritional deficiencies, 4, 14, 36, 41, 63, 70, 133, 143, 187, 208–13
 nutritional requirements, 13, 22, 24, 29, 155, 203
 overnutrition, 30, 133–34, 144–45, 248, glossary
 undernutrition, 6, 22, 71, 133–35,
137, 144–45, 147, 168, 180, 248, glossary
 See also Stress, nutritional
Obesity, 4, 6–7, 13–14, 21, 24–25, 28, 30, 110, 123, 133–34, 144–45, chapter 10, 243, 247–48, glossary
 International Obesity Task Force (IOTF), 225
 See also Adiposity; Nutrition, overnutrition
Omnivore, 29, 64, 242, 247
Osteoporosis, 7, chapter 9, 243
 fractures and, 7, chapter 9

Palaeopathology, 41, 188, 205–6, 212–13
 cribra orbitalia, 40–41, 51, 89, 209, glossary
 nonspecific indicators of stress, 40–41, 51, 89, 178, 209
 porotic hyperostosis, 40, glossary
 rickets, 41, 185, 211, glossary
Paleolithic diet, 13–14, 242
Pastoralism, 21, 26–27, 37, 50
 See also Herding
Physical activity, 21, 25–29, 40, 63, 70–71, 134, 144–45, 197–204, 208, 213, 223, 236
 Physical Activity Level (PAL), 25–27, glossary
Pleistocene, 19, 37
Political-economy, 1–7, 117, 152, 242, 244
Population Attributable Risk (PAR%), 233–34
Poverty, 136, 152–55, 158, 161, 164, 166, 168, 180
 See also Development (economic); Food security
Primates, 14–20, 29, 60–62, 66, 228, 245–46
 captive, 61–63, 70
 diet and nutrition, 16–17, 20, 29, 62–63, 67, 245
 great apes, 16–17, 20, 61–65, 71–72, 75, 245
 lactation biology, 4, 57–73

New World monkeys, 16
Old World monkeys, 62–63, 69–70

Ranging, 4, 14, 18–21, 35, 38
See also Foraging
Reproduction, 4, 35, 59, 177, 200–1, 204, 208, 210, 228, 246
  fetal development, 43, 63–64, 94, 99, 113, 179, 202–4
  pregnancy, 7, 35, 63–65, 70–72, 113, 116, 198, 202–3, 210, 228
  See also Birth
Roman period, 5, 7, chapters 4 and 8, 244
Rural, 6, 69, 133, chapter 6, 161
  See also Migration

Seasonality, 35, 38, 71, 92, 137–38, 143, 145, 153, 228, 245, glossary
Sedentism, 26, 30, 50, 206, 223, 228
Sexual dimorphism, 73, 210
Skeleton
  body mass and, 38
  bone strength, 37
  cranial growth, 19, 40–45, 51
  postcranial growth, 7, 38, 40–46, 51, 174, 180, 184, 188, 198
  See also Age, dental and skeletal age estimates; Aging; Bone; Osteoporosis
Socioeconomic status (SES), 115, 118–19, 124–26, 134, 204, 209, glossary
South America, 24, 206, 231
  Bolivia, 22–23, 27
  Ecuador, 23, 27, 145
Stable isotope analysis, 4, 38–39, chapter 4, 182, 185–87, 247
  carnivore effect, 91, 96, 98, 101

diet and, 3, 5, 57, 38–39, 50, 69, chapter 4, 174, 181–88, 247
IYCF and, chapters 4 and 8
trophic levels, 38, 90–91, 94–96, 101, 103, 181, 185–86, 247, glossary
Stone Age, 4, chapter 2, 243
Stress
  developmental, 6, 180, 184–85, 246
  environmental, 93, 209–10
  nutritional, 7, 22, 63, 206, 209
  pregnancy/lactation and, 210
  stressors, 179–80, 188
  weaning and, 50
  See also Palaeopathology, nonspecific indicators of stress
Subsistence, 2, 21, 25–30, 51, 69, 135, 145, 161, 242, 245, 247

UN Human Development Index, 153
UNICEF, 110
Urban, 6, 25, 69, chapter 6, 165, 244
  urbanization and nutrition, chapter 6
  See also Migration

Weaning 4–5, 39, 41, 50, 58–75, 89–92, 94, 98–103, 140–41, 155, 162–64, 176–78, 181, 187–88, 203–4, 242, 244, 246, glossary
  weaning patterns, 4, 5, 94
  See also Breastfeeding; IYCF; Primates, lactation biology
Weight-for-height
  See Growth and development, measures of
World Bank, 158
World Health Organization (WHO), 25, 39, 111, 155–56

RECEIVED

JUL 2 0 2012

GUELPH HUMBER LIBRARY
205 Humber College Blvd
Toronto, ON   M9W 5L7

Due Date	Date Returned
T/OCT 29, 12	DEC 2 0 2012
T/Feb 07, 13	APR 1 2 2013
APR 2 0 2014	DEC 1 7 2014
T/OCT 06, 15	
NOV 0 6 2015	NOV 1 0 2015
www.library.humber.ca	